SUMMARY. We study the actions of discrete amenable groups on factor von Neumann algebras. We give the classification up to outer conjugacy of the actions of amenable groups on the type II hyperfinite factors. A main result is the unicity up to outer conjugacy of the free action of an amenable group on the hyperfinite type II_1 factor.

ACKNOWLEDGMENTS. The results of this paper were mainly obtained during my activity as a research fellow at INCREST-Bucharest, Romania, and were announced in [34]. The writing up, as well as several improvements in the proof, were done at the University of Warwick, England.

I would like to thank Ciprian Foiaş, as well as my INCREST colleagues Grigore Arsene, Zoia Ceauşescu, Mihai Pimsner, Sorin Popa, Dan Voiculescu, and especially Şerban Strătilă for their support. I received further support from David Evans, Klaus Schmidt and Christopher Zeeman at Warwick.

Benjamin Weiss was very kind to send me a description of the results in [36] prior to their publication.

I am grateful to Vaughan Jones for useful discussions and for the generosity with which he helped make the results in this paper known.

My wife Ileana generously helped me carry on my work. Deborah Craig typed and edited the manuscript with remarkable skill and accuracy.

Lecture Notes in Mathematics

Edited by A. Dold and B. Eckmann

1138

Adrian Ocneanu

Actions of Discrete Amenable Groups on von Neumann Algebras

Springer-Verlag
Berlin Heidelberg New York Tokyo

Author

Adrian Ocneanu
Department of Mathematics, University of California
Berkeley, California 94720, USA

Mathematics Subject Classification (1980): 20F29, 46L40, 46L55

ISBN 3-540-15663-1 Springer-Verlag Berlin Heidelberg New York Tokyo
ISBN 0-387-15663-1 Springer-Verlag New York Heidelberg Berlin Tokyo

Printing and binding: Beltz Offsetdruck, Hemsbach/Bergstr.
2146/3140-543210

TABLE OF CONTENTS

INTRODUCTION

In this paper we study automorphic actions of discrete groups on von Neumann algebras. The main result is the following.

THEOREM. *Let* G *be a countable discrete amenable group and let* R *be the hyperfinite* II_1 *factor. Any two free actions of* G *on* R *are outer conjugate.*

An action α of G on a factor M is a homomorphism of G into Aut M, the group of automorphisms of M; α is called free if α_g is not inner for any $g \in G$, $g \neq 1$. Two actions $\alpha, \bar{\alpha}$: $G \longrightarrow \text{Aut } M$ are called outer conjugate if there exists a unitary cocycle u for , i.e. unitaries $u_g \in M$, $g \in G$, with

$$u_{gh} = u_g \alpha_g (u_h)$$

and $\theta \in \text{Aut } M$ such that

$$\bar{\alpha}_g = \theta \text{ Ad } u_g \alpha_g \theta^{-1} , \quad g \in G .$$

The amenability restriction is essential: for any nonamenable group G, the above theorem does not hold [26]. Actions of general amenable groups arise naturally in connection with hyperfinite factors [27, Theorem 3.1].

We actually work with more general factors and actions. We only require the factor M to be isomorphic to $M \otimes R$ and to have separable predual. For such a factor, we prove the outer conjugacy for actions which are centrally free (i.e. each α_g, $g \neq 1$, acts non-trivially on central sequences) and approximately inner (i.e. each α_g is a limit of inner automorphisms).

For a (not necessarily free) action α of a discrete amenable group G on R, we show that the characteristic invariant $\Lambda(\alpha)$, introduced in [21], is complete for outer conjugacy. On the hyperfinite II_∞ factor $R_{0,1}$, the system of invariants $(\Lambda(\alpha), \text{mod}(\alpha))$ is complete for outer conjugacy, where mod: $\text{Aut } R_{0,1} \longrightarrow \mathbb{R}_+$ is the module ([4]).

It seems possible to go along the lines of [6] to obtain the classification for type III factors as well.

We do a parallel study of G-kernels on factors, which are homomorphisms of the group G into Out M = Aut M/Int M. Two G-kernels $\beta, \bar{\beta}$ are conjugate if there exists $\theta \in \text{Out } M$ with

$$\beta_g = \theta \beta_g \theta^{-1} , \quad g \in G .$$

We show that, if G is a discrete amenable group, and for a

G-kernel β on R, the Eilenberg-MacLane H^3-obstruction $Ob(\beta)$ is a complete conjugacy invariant, and for a G-kernel β on $R_{0,1}$, $(Ob(\beta), mod(\beta))$ is a complete system of invariants to conjugacy.

A result of independent interest obtained is the vanishing of the 2-dimensional unitary valued cohomology for centrally free actions (the 1-cohomology does not vanish for infinite groups: there are many examples of outer conjugate but not conjugate actions).

Involutory automorphisms of factors have been studied by Davies [8], but the major breakthrough was done by Connes in [3], where he classified the actions of $\mathbb{Z}_n$ on R, and in [4], where he classified actions of $\mathbb{Z}$ up to outer conjugacy. A study of the cohomological invariants for group actions was done by Jones in [21] where he extended the characteristic invariant of [3] to group actions. In [23] Jones classified the actions of finite groups on R, up to conjugacy. Product type actions of $\mathbb{Z}_n$ of UHF algebras were classified by Fack and Marechal [11], and Kishimoto [27], and finite group actions on C^*-algebras were studied by Rieffel [39]. Classification results for finite group actions on AF-algebras were obtained in [17],[18] by Herman and Jones.

This paper is an extension of [4], and also generalizes the outer conjugacy part of [23].

In the first chapter we state the main results in their general setting, and in the second chapter we use them to obtain, in the presence of invariants, classification results on the hyperfinite type II factors. The proofs of the main results are done in the remaining part of the paper.

The first problem is to reduce the study of the group G to one of its finite subsets. An approximate substitute for a finite G-space is an almost invariant finite subset of G, obtained from amenability by means of the Følner Theorem. A link between such subsets is yielded by the Ornstein and Weiss Paving Theorem. By means of a repeated use of these procedures we obtain a Paving Structure for G, which is a projective system of finite subsets of G, endowed with an approximate G-action. We use this structure to construct a faithful representation of G on the hyperfinite II_1 factor, well provided with approximations on finite dimensional subfactors.

The main ingredients of the construction are the Mean Ergodic Theorem applied on the limit space of the Paving Structure, together with a combinatorial construction of multiplicity sets. We call the inner action yielded by this representation the submodel action. A tensor product of countably many copies of the submodel action is used as the model of free action of G. For $G = \mathbb{Z}$ this model is different from the one used in [4].

An essential feature of Connes' approach is the study of automorphisms in the framework of the centralizing ultraproduct algebra M_ω, introduced by Dixmier and McDuff. In the fifth chapter we make a systematic study of these techniques and also introduce the normalizing algebra M^ω as a device for working with both the algebra M and the centralizing algebra M_ω.

We continue with the main technical result of the paper, the Rohlin Theorem, which yields, for centrally free actions of amenable groups, an equivariant partition of the unit into projections. In the first part of the proof we obtain some, possibly small, equivariant system of of projection. The approach is based on the study of the geometry of the crossed product, and makes use of a result of S.Popa on conditional expectations in finite factors [37]. In the second part we put together such systems of projections to obtain a partition of unity. We use a procedure in which at each step the construction done in the previous steps is slightly perturbed. These methods yield new proofs of the Rohlin Theorem both for amenable group actions on measure spaces and for centrally free actions of $\mathbb{Z}$ on von Neumann algebras.

As a consequence of the Rohlin Theorem, we obtain in the seventh chapter stability properties for centrally free actions of amenable groups. We first prove an approximate vanishing of the one- and two-dimensional cohomology. The main stability result is the exact vanishing of the 2-cohomology. The proof is based on the fact that in any cohomology class there is a cocycle with an approximate periodicity property with respect to the previously introduced Paving Structure. The techniques used here yield an alternative approach for the study of the 2-cohomology on measure spaces. The usual way is to reduce the problem, by means of the hyperfiniteness, to the case of a single automorphism, where the 2-cohomology is always trivial.

The final part of the paper deals with the recovery of the model inside given actions. We first show that there are many systems of matrix units approximately fixed by the action. From such a system, together with an approximately equivariant system of projections given by the Rohlin Theorem, we obtain an approximately equivariant system of matrix units; this is precisely how a finite-dimensional approximation of the submodel looks. Repeating the procedure we obtain an infinite number of copies of the submodel and thus a copy of the model. At each of the steps of this construction there appear unitary perturbations. The vanishing of the 2-cohomology permits the reduction of those perturbations arbitrarily close to 1 cocycles.

The corresponding results for G-kernels are obtained by removing from the proofs the parts connected to the 2-cohomology vanishing.

The last chapter contains the proof of the Isomorphism Theorem. Under the supplementary assumption that the action is approximately inner we infer that on the relative commutant of the copy of the model that we construct, the action is trivial; i.e. the model contains the whole action. We begin by obtaining a global form from the elementwise definition of approximate innerness. Approximately inner automorphisms are induced by unitaries in the ultraproduct algebra M^{ω}. We use a technique of V.Jones to work, by means of an action of $G \times G$, simultaneously with these unitaries and with the action itself. After constructing, in the same way as in the preceding chapter, an approximately equivariant system of matrix units, we make it contain the unitaries that approximate the action. We obtain a copy of the submodel which contains a large part of the action, in the sense that for many normal states on M, the restriction to the relative commutant of the copy of the submodel is almost fixed by the action. This way of dealing with the states of the algebra, in view of obtaining tensor product splitting of the copy of the model, is different from the one in [4], and avoids the use of spectral techniques.

A characteristic of the framework of this paper is the superposition at each step of technical difficulties coming from the structure of general amenable groups, and from the absence of a trace on the factor. Nevertheless, in a technically simple context like, e.g. $\mathbb{Z}^2$ acting on R, all the main arguments are still needed.

With techniques based on the Takesaki duality, V.Jones [24] obtained from the above results the classification of a large class of actions of compact abelian groups (the duals of which are discrete abelian, hence amenable, groups).

A similar approach towards classifying actions of compact non-abelian groups would first require a study of the actions of their duals, which are precisely the discrete symmetrical Kac algebras. A natural framework for this extension is the one of discrete amenable Kac algebras, which includes both the duals of compact groups and the discrete amenable groups. It appears [35] that such an approach can be done along lines similar to the ones in this paper. A first step is to provide, in the group case, proofs which are of a global nature, i.e. deal with subsets rather than with elements of the group; the proof of the Rohlin Theorem given in this paper is such an instance. Apart from that, the subsequent extension to the non-groupal case needs, in general, techniques having no equivalent in the group case.

NOTATION

Let M be a von Neumann algebra. M^h, M^+, M_1, $Z(M)$, $U(M)$, Proj M denote the hermitean part, positive part, unit ball, center, unitary group, and projection lattice of M, respectively. M_* and M_*^+ denote the predual of M and its positive part.

If $\phi \in M_*$ and $x,y \in M$, then $\phi x, x\phi \in M_*$ are defined by $(\phi x)(y) = \phi(xy)$; $(x\phi)(y) = \phi(yx)$. We let $[\phi,x] = \phi x - x\phi$.

If $\phi \in M_*^+$ and $x \in M$, we let $|x|_\phi = \phi(|x|)$, $\|x\|_\phi = \phi(x^*x)^{\frac{1}{2}}$, and $\|x\|_\phi^\# = \phi(\frac{1}{2}(x^*x + xx^*))^{\frac{1}{2}}$.

Chapter 1: MAIN RESULTS

This chapter contains an outline of the results of independent interest obtained in the main body of the paper.

1.1 Let M be a von Neumann algebra. An automorphism θ of M is called centrally trivial, $\theta \in \mathrm{Ct}M$, if for any centralizing sequence $(x_n) \in M$, i.e. which is norm bounded and satisfies $\lim_{n\to\infty} \|[\phi,x_n]\| = 0$ for any $\phi \in M_*$, one has $\theta(x_n)-x_n \longrightarrow 0$ *-strongly. θ is called properly centrally nontrivial if $\theta|pM$ is not centrally trivial for any nonzero θ-invariant projection p in $Z(M)$. A discrete group action $\alpha: G \to \mathrm{Aut}\, M$ is called centrally free if for any $g \in G\backslash\{1\}$, α_g is properly centrally nontrivial.

The group G dealt with in this section will always be assumed countable and discrete.

A cocycle crossed action of the group G on M is a pair (α,u), where $\alpha: G \to \mathrm{Aut}\, M$ and $u: G \times G \to U(M)$ satisfy for $g,h,k \in G$

$$\alpha_g\alpha_h = \mathrm{Ad}u_{g,h}\alpha_{gh} \quad ,$$

$$u_{g,h}\, u_{gh,k} = \alpha_g(u_{h,k})\, u_{g,hk} \quad ,$$

$$u_{1,g} = u_{g,1} = 1 \quad .$$

(α,u) is called centrally free if α is free with the obvious adaptation of the definition. The cocycle u is the coboundary of v, $u = \partial v$, if $v: G \to U(M)$ satisfies

$$u_{g,h} = \alpha_g(v_h^*)v_g^*\, v_{gh} \quad .$$

In this case (α,u) may be viewed as a perturbation of the action $(\mathrm{Ad}\, v_g\alpha_g)$. We shall prove in Chapter 7 the following vanishing result for the 2-cohomology.

THEOREM. *Let* G *be an amenable group, let* M *be a von Neumann algebra with separable predual, and let* $\phi \in M_*^+$ *be faithful. If* (α, u) *is a centrally free cocycle crossed action of* G *on* M*, such that* $\alpha|Z(M)$ *preserves* $\phi|Z(M)$*, then* u *is a coboundary.*

Moreover, given any $\varepsilon > 0$ *and any finite* $F \subset G$*, there exists* $\delta > 0$ *and a finite* $K \subset G$ *such that if*

$$\|u_{g,h} - 1\|_\phi^\# < \delta \qquad g,h \in K$$

then $u = \partial v$ *with*

$$\|v_g - 1\|_\phi^\# < \varepsilon \qquad g \in F .$$

A similar result for the 1-cohomology holds only if G is finite, in which case the classification can be carried on up to conjugacy [23].

1.2 A factor M is called a McDuff factor if it is isomorphic to $R \otimes M$, where R is the hyperfinite II_1 factor. Several equivalent properties, due to McDuff and Connes are given in **5.2** below.

In **8.5** we shall obtain the following result.

THEOREM. *Let* G *be an amenable group and let* M *be a McDuff factor with separable predual. If* $\alpha: G \longrightarrow \mathrm{Aut}\, M$ *is a centrally free action then* α *is outer conjugate to* $id_R \otimes \alpha$.

Moreover, given any $\varepsilon > 0$*, any finite* $K \subset G$*, and any* $\phi \in M_*^+$*, there exists an* (α_g)*-cocycle* (v_g) *such that* $(\mathrm{Ad}\, v_g \alpha_g)$ *is conjugate to* $id_R \otimes \alpha$ *and*

$$\|v_g - 1\|_\phi^\# \qquad g \in K .$$

Actually, the central freedom of α is basically used only to obtain cocycles. An alternative approach based on Lemma 2.4 would not need this assumption.

1.3 In Chapter 4 we construct a model of free action $\alpha^{(0)}: G \to \mathrm{Aut}\, R$ for an amenable group G. In **8.6** we show that this model action is contained in any centrally free action.

THEOREM. *Let* G *be an amenable group and let* M *be a McDuff factor with separable predual. Any centrally free action* $\alpha: G \to \mathrm{Aut}\, M$ *is outer conjugate to* $\alpha^{(0)} \otimes \alpha$.

Moreover, as in the preceding theorem, the cocycle that appears can be chosen arbitrarily close to 1.

1.4 Under the supplementary assumption that each α_g is approximately inner, the action is shown in **9.3** to be uniquely determined up to outer conjugacy.

THEOREM. *Let* G *be an amenable group and let* M *be a McDuff factor with separable predual. Any centrally free approximately inner action* $\alpha: G \to \mathrm{Aut}\, M$ *is outer conjugate to* $\alpha^{(0)} \otimes \mathrm{id}_M$.

Bounds on the cocycle may also be obtained.

COROLLARY. *Any two free actions of the amenable group* G *on* R *are outer conjugate.*

Proof. By results of Connes [3], $\mathrm{Ct}R = \mathrm{Int}\, R$ and $\overline{\mathrm{Int}}\, R = \mathrm{Aut}\, R$.

1.5 The study of actions of groups is closely connected to the study of G-kernels, which are homomorphisms $G \to \mathrm{Out}\, M = \mathrm{Aut}\, M/\mathrm{Int}\, M$. Since inner automorphisms are centrally trivial, central freedom can be defined for G-kernels. From the proof of Theorem 1.2 in **8.8** we obtain the analogous result for G-kernels.

THEOREM. *Let* G *be an amenable group and* M *a McDuff factor with separable predual. Any centrally free G-kernel* $\beta: G \to \mathrm{Out}\, M$ *is conjugate to* $\mathrm{id}_R \otimes \beta$.

1.6 In the same way we obtain in **8.9** the following analogue of Theorem 1.3.

THEOREM. *Let* G *be an amenable group and* M *a McDuff factor with separable predual. Any centrally free G-kernel* $\beta: G \to \mathrm{Out}\, M$ *is conjugate to* $\Pi(\alpha^{(0)}) \otimes \beta$.

Here $\alpha^{(0)}: G \to \mathrm{Aut}\, R$ is the model action and $\Pi: \mathrm{Aut}\, M \to \mathrm{Out}\, M$ is the canonical projection.

Chapter 2: INVARIANTS AND CLASSIFICATION

We obtain the outer conjugacy classification of amenable group actions on the type II_1 and II_∞ hyperfinite factors from the results in the preceding chapter.

2.1 When an action has an inner part, there appears a cohomological invariant coming from the uniqueness modulo a scalar of the unitaries implementing it. This invariant, called the characteristic invariant, introduced by Connes for actions of $\mathbb{Z}_n$ in [3], was defined for general discrete groups by Jones [21]. We shall briefly describe it in what follows.

Let α be an action of a discrete group G on a factor M. A first conjugacy invariant is the normal subgroup $N(\alpha) = \alpha^{-1}(\mathrm{Int}\, M)$ of G. For each h $N = N(\alpha)$, we choose a unitary $v_h \in N$ such that $\alpha_h = \mathrm{Ad}\, v_h$ and take $v_1 = 1$. For $h,k \in N$, both $v_h v_k$ and v_{hk} implement $\alpha_h \alpha_k = \alpha_{hk}$, and thus there exists $\mu_{h,k} \in \mathbb{T} = \{z \in \mathbb{C} \mid |z|=1\}$ such that

$$v_h v_k = \mu_{h,k}\, v_{hk} \quad .$$

Similarly for $g \in G$ and $h \in N$, since $\alpha_g \alpha_{g^{-1}hg} \alpha_{g^{-1}} = \alpha_h$, we infer

$$\alpha_g(v_{g^{-1}hg}) = \lambda_{g,h}\, v_h \quad \text{for some} \quad \lambda_{g,h} \in \mathbb{T} \, .$$

The pair (λ,μ) of maps $\lambda\colon G \times N \to \mathbb{T}$, $\mu\colon N \times N \to \mathbb{T}$ satisfies the following relations for $h,k,\ell \in N$, $g,j \in G$:

$$\mu_{h,k}\mu_{hk,\ell} = \mu_{k,\ell}\, \mu_{h,k\ell}$$

$$\lambda_{gf,h} = \lambda_{g,h}\, \lambda_{f,g^{-1}hg}$$

$$\lambda_{h,k} = \mu_{h,h^{-1}kh}\, \mu^*_{k,h}$$

$$\lambda_{g,hk}\lambda^*_{g,h}\lambda^*_{g,k} = \mu_{h,k}\, \mu^*_{g^{-1}hg,g^{-1}kg}$$

$$\lambda_{g,1} = \lambda_{1,h} = \lambda_{h,1} = \lambda_{1,h} = 1$$

where * denotes the complex conjugation. This follows by easy computation from the definitions of λ and μ. We let $Z(G,N)$ be the abelian group consisting of all the pairs of functions (λ,μ) satisfying the above relations.

To get rid of the dependence of (λ,μ) on the choice of (v_h), we let $C(N)$ be the set of all maps $\eta\colon N \to \mathbb{T}$ with $\eta_1 = 1$ and, for

$\eta \in C(N)$, we let $\partial\eta = (\lambda,\mu)$ where

$$\lambda_{g,h} = \eta_h \eta_{g^{-1}hg}$$

$$\mu_{h,k} = \eta_{hk}\eta_h^*\eta_k^* \qquad g \in G\ , \quad h,k \in H\ .$$

It is easy to see that $B(G,N) = \partial C(N)$ is a subgroup of $Z(G,N)$; we denote by $\Lambda(G,N)$ the quotient $Z(G,N)/B(G,N)$. For an action α, the image $\Lambda(\alpha) = [\lambda,\alpha]$ of (λ,α) in $\Lambda(G,N)$ no longer depends on the choice of the unitaries (v_g) and hence is a conjugacy invariant. If (w_g) is an (α_g)-cocycle and $\tilde{\alpha}_g = \mathrm{Ad}\, w_g\alpha_g$, then for $h \in N$, $\tilde{v}_h = w_h v_h$ implements $\tilde{\alpha}_h$, and it is easy to compute that these unitaries yield the same pair (λ,μ) for $\tilde{\alpha}$. Thus $\Lambda(\alpha)$ is an outer conjugacy invariant, called the characteristic invariant of the action.

When N is abelian, then $[\lambda,\mu]$ depends only on λ and no quotient has to be taken.

The characteristic invariant can also be defined in terms of group extensions. Let $\alpha\colon G \longrightarrow \mathrm{Aut}\, M$ with $N = \alpha^{-1}(\mathrm{Int}\, M)$ and let $\tilde{N} = \{(h,u) \in N \times U(M) \mid \alpha_h = \mathrm{Ad}u\}$. Then $\tilde{N}$ is a subgroup of $N \times U(N)$ and the maps $\mathbb{T} \to \tilde{N}$, $t \to (1,t)$ and $\tilde{N} \to N\colon (h,u) \to h$ yield an exact sequence

$$1 \longrightarrow \mathbb{T} \longrightarrow \tilde{N} \longrightarrow N \longrightarrow 1$$

where the induced action of N on $\mathbb{T}$ by conjugation is trivial. Moreover, $g \in G$ acts on N by conjugation: $h \longrightarrow ghg^{-1}$, and if we let it act on $\mathbb{T}$ trivially and on $\tilde{N}$ by $(h,u) \longrightarrow (ghg^{-1}, \alpha_g(u))$, the above sequence becomes an exact sequence of G-modules.

One can show that the classes of extensions of N by $\mathbb{T}$ (trivial action) in the category of G-modules form a group with the Brauer product and this group is naturally isomorphic to $\Lambda(G,N)$.

2.2 Cohomological invariants for the conjugacy of G-kernels were defined in an algebraic context by Eilenberg and McLane and adapted to von Neumann algebras by Nakamura and Takeda [32] and Sutherland [43].

Let $\beta\colon G \longrightarrow \mathrm{Out}\, M$ be a G-kernel on a factor and let $\alpha\colon G \longrightarrow \mathrm{Aut}\, M$ be a section of it, with $\alpha_1 = 1$. For each $g,h \in G$, there are unitaries $w_{g,h} \in M$ with

$$\alpha_g\alpha_h = \mathrm{Ad}\, w_{g,h}\alpha_{gh}$$

which may be assumed to satisfy $w_{1,g} = w_{g,1} = 1$. From the associativity relation $(\alpha_g\alpha_h)\alpha_k = \alpha_g(\alpha_h\alpha_k)$ one obtains

$$w_{g,h}w_{gh,k} = \delta_{g,h,k}\,\alpha_g(w_{h,k})\, w_{g,hk}$$

for some $\delta_{g,h,k} \in \mathbb{T}$. The function $\delta: G^3 \to \mathbb{T}$ satisifes a normalized 3-cocycle relation, and its class $Ob(\beta)$ in $H^3(G,\mathbb{T})$, called the obstruction, is a conjugacy invariant for the G-kernel β.

Jones has shown that if G is a countable discrete group and if R is the hyperfinite II_1 factor, then for any normal subgroup N of G and any $[\lambda,\mu] \in \Lambda(G,N)$ there exists an action $\alpha: G \to \mathrm{Aut}\, R$ with $N(\alpha) = N$ and $\Lambda(\alpha) = [\lambda,\mu]$, and for each $[\delta] \in H^3(G)$ there exists a free G-kernel $\beta: G \to \mathrm{Out}\, R$ with $Ob(\beta) = [\delta]$.

Let N be a normal subgroup of G and let $Q = G/H$. One can define natural connecting maps to extend the Hochschild-Serre exact sequence to an eight-term exact sequence

$$1 \to H^1(Q) \to H^1(G) \to H^1(N)^G \to H^2(Q) \to H^2(G) \to \Lambda(G,N) \to H^3(Q) \to H^3(G) \quad .$$

For details see [19],[22],[38].

2.3 The following lemma describes actions with trivial characteristic invariant.

LEMMA. *Let* G *be a countable discrete amenable group and let* M *be a factor with separable predual. Let* $\alpha: G \to \mathrm{Aut}\, M$ *be an action with* $\alpha^{-1}(\mathrm{Int}\, M) = \alpha^{-1}(\mathrm{Ct}\, M) = N$. *Let* $p: G \to Q = G/N$ *be the canonical projection. If* $\Lambda(\alpha)$ *is trivial then there exist an* α*-cocycle* u *and an action* $\tilde{\alpha}: Q \to \mathrm{Aut}\, M$ *such that*

$$\mathrm{Ad}\, u_g \alpha_g = \tilde{\alpha}_{p(g)} , \qquad g \in G .$$

Proof. By the triviality of $\Lambda(\alpha)$, we may choose a map $v: N \to U(M)$, $v_1 = 1$ such that for $g \in G$, $h,k \in N$ we have

$$\alpha_h = \mathrm{Ad}\, v_h , \qquad v_h v_k = v_{hk} , \qquad \alpha_g(v_{g^{-1}hg}) = v_h \quad .$$

Let $s: Q \to G$ be a section of p with $s(1) = 1$ and let $\bar{\alpha}_q = \alpha_{s(q)}$ for $q \in Q$. If $q,r \in Q$, define $t(q,r) \in N$ by

$$s(q)s(r) = t(q,r)s(qr)$$

and let $\bar{w}_{q,r} = v_{t(q,r)}$. We have for $q,r,s \in Q$

$$t(q,r)t(qr,s) = \mathrm{Ad}(s(q))(t(r,s))t(q,rs)$$

hence $((\bar{\alpha}_q),(\bar{w}_{q,r}))$ is a cocycle crossed action of Q on M, which is, by the hypothesis of the lemma, centrally free. The 2-cohomology vanishing (Theorem 1.1) yields a map $z: Q \to U(M)$, $z_1 = 1$, with

$$z_q\bar{\alpha}_q(z_r)\bar{w}_{q,r}z_{q,r}^* = 1 \qquad q,r \in Q .$$

Let $\tilde{\alpha}_q = \mathrm{Ad}\, z_q\bar{\alpha}_q$. Then $\tilde{\alpha}$ is an action of Q on R. For $g \in G$ with $p(g) = p$ and $g = hm$, $h \in H$, $m = s(p)$, we let $u_g = z_p v_h^*$. We have

$$\mathrm{Ad}\, u_g\alpha_g = \mathrm{Ad}(z_p v_h^*)\ \mathrm{Ad}\, v_h\bar{\alpha}_p = \tilde{\alpha}_p$$

and all that remains to be shown is that (u_g) is an α-cocycle.

Let $g,f \in G$; $p = p(g)$, $q = p(f)$ Q; $m = s(p)$, $n = s(q)$, $r = s(pq) \in S(q) \subseteq G$; $h = gm^{-1}$, $k = fn^{-1}$, $\ell = gfr^{-1} \in N$. We have

$$t(p,q) = mnr^{-1} = mnf^{-1}g^{-1}\ell = mk^{-1}m^{-1}h^{-1}\ell = \mathrm{Ad}(s(p))(k^{-1})h^{-1}\ell$$

so that

$$\bar{w}_{p,q} = \alpha_p(v_k^*)v_h^*v_\ell$$

and we obtain

$$u_g\alpha_g(u_f)u_{gf}^* = z_p v_h^* v_h \bar{\alpha}_p(z_q v_k)v_h^* v_\ell z_{pq}^* = z_p\bar{\alpha}_p(z_q)\bar{w}_{p,q}z_{pq}^* = 1 .$$

The lemma is proved.

2.4 The lemma that follows is a device to obtain cocycles, inspired by [22].

LEMMA. *Let* M,N,P *be factors and let* $\alpha: G \to \mathrm{Aut}\, M$, $\beta: G \to \mathrm{Aut}\, N$ *be actions of a discrete group* G. *Let* $\gamma: G \to \mathrm{Aut}\, P$ *and* $v: G \to U(M)$ *be maps such that*

β *is conjugate to* $\beta \otimes \beta$,

$(\mathrm{Ad}\, v_g\alpha_g)$ *is conjugate to* $\beta \otimes \gamma$.

Then there exists an α *cocycle* u *such that*

$(\mathrm{Ad}\, u_g\alpha_g)$ *is conjugate to* $\beta \otimes \alpha$.

Proof. Since $(\mathrm{Ad}\, v_g\alpha_g)$ is conjugate to $\beta \otimes \gamma$ and to $\beta \otimes \beta \otimes \gamma$, there exists an isomorphism $\theta: N \otimes M \to M$ such that

$$\mathrm{Ad}\, v_g\alpha_g = \theta(\beta_g \otimes \mathrm{Ad}\, v_g\alpha_g)\theta^{-1} .$$

Let $\bar{v}_g = \theta(1_N \otimes v_g^*)v_g$; then

$$\mathrm{Ad}\, \bar{v}_g\alpha_g = \theta(\beta_g \otimes \alpha_g)\theta^{-1} .$$

The right member is an action, hence

$$z_{g,h} = \bar{v}_g\alpha_g(\bar{v}_h)\bar{v}_{gh}^*$$

is a scalar for $g,h \in G$.

Once again, since β is conjugate to $\beta \otimes \beta$, there exists an isomorphism $\bar{\theta}: N \otimes M \to M$ such that

$$\mathrm{Ad}\, \bar{v}_g \alpha_g = \bar{\theta}(\beta_g \otimes \mathrm{Ad}\, \bar{v}_g \alpha_g)\bar{\theta}^{-1} .$$

We let $u_g = \bar{\theta}(1 \otimes \bar{v}_g^*)\bar{v}_g$ and infer

$$\mathrm{Ad}\, u_g \alpha_g = \bar{\theta}(\beta_g \otimes \alpha_g)\bar{\theta}^{-1} ,$$

$$\begin{aligned} u_g \alpha_g(u_h) u_{gh}^* &= \theta(1 \otimes \bar{v}_g^*)(\mathrm{Ad}\, \bar{v}_g \alpha_g)(1 \otimes v_h^*)\bar{v}_g \alpha_g(\bar{v}_h)\bar{v}_{gh}^* \bar{\theta}(1 \otimes \bar{v}_{gh}) \\ &= z_{g,h}\ \bar{\theta}(1 \otimes (\bar{v}_g^*(\mathrm{Ad}\, \bar{v}_g \alpha_g)(\bar{v}_h^*)\bar{v}_{gh})) \\ &= z_{g,h} z_{g,h}^* = 1 . \end{aligned}$$

The lemma is proved.

2.5 The preceding chapter contained classification results in the invariantless case. In many situations, we can reduce to this case by tensoring with model actions having opposite invariants. The formal setting is the following.

LEMMA. *Let Γ be a discrete group.*

a) *Let Σ be a unital semigroup and let $\psi: \Sigma \to \Gamma$ be a surjective homomorphism with $\phi^{-1}(1) = \{1\}$. Then ψ is bijective.*

b) *Let Δ be a (left) Γ-space and let $\phi: \Delta \to \Gamma$ be a homomorphism of Γ-spaces with* card $\phi^{-1}(1) = 1$. *Then ϕ is bijective.*

Proof. a) For $x,y \in \Sigma$ with $\psi(x) = \psi(y) = g$, we find $z \in \Sigma$ with $\psi(z) = g^{-1}$. Then $\psi(xz) = \psi(zy) = 1$ and so $xz = zy = 1$, hence

$$x = x.1 = xzy = 1.y = y .$$

b) Let $\Gamma \times \Delta \to \Delta$, $(g,x) \to gx$ be the action of Γ on Δ, and let $e = \phi^{-1}(1) \in \Delta$. For any $x \in \Sigma$ we have $\phi(\phi(x)^{-1}x) = 1$, so $\phi(x)^{-1}x = e$ and

$$x = 1.x = \phi(x)\phi(x)^{-1}x = \phi(x)e .$$

Hence $g \to ge: \Gamma \to \Delta$ is the inverse map of ϕ.

2.6 We begin by classifying actions on the hyperfinite II_1 factor R.

THEOREM. *Let G be a countable discrete amenable group. Two* *actions* $\alpha,\beta\colon G \to \mathrm{Aut}\, R$ *are outer conjugate if and only if* $N(\alpha)=N(\beta)$ *and* $\Lambda(\alpha)=\Lambda(\beta)$.

Proof. We keep a normal subgroup N of G fixed; let Γ be the group $\Lambda(G,N)$ and let Σ be the set of outer conjugacy classes $[\alpha]$ of actions $\alpha\colon G \to \mathrm{Aut}\,\bar{R}$, with $\bar{R}$ isomorphic to R and $N(\alpha)=N$. We let $\psi\colon \Sigma \to \Gamma$ be the characteristic invariant, which is well defined on outer conjugacy classes. Σ is a semigroup with tensor product multiplication, which preserves classes, and ψ is a semigroup morphism; by the results of Jones ψ is surjective. To apply Lemma 2.5(a) it remains to show that Σ is unital and $\psi^{-1}(1)=\{1\}$. By Lemma 2.3, in the class of any action $\alpha\colon G \to \mathrm{Aut}\, R$ with $\psi([\alpha])=1$ there is an action $\bar{\alpha}$ induced by a free action $\tilde{\alpha}$ of the quotient $Q=G/H$. Since Q is amenable, any two such actions of Q are outer conjugate by Corollary 1.4, and the cocycle lifts to a cocycle of $\bar{\alpha}$. Hence the preimage of $1\in\Gamma$ consists of a single class.

Let $\tilde{\alpha}\colon Q \to \mathrm{Aut}\, R$ be the model of free action and let $\alpha\colon G \to \mathrm{Aut}\, R$ be the induced G-action. Let $\beta\colon G \to \mathrm{Aut}\, R$ with $[\beta]\in\Sigma$; then β induces a Q-kernel $\beta'\colon G \to \mathrm{Out}\, R$. If $\Pi\colon \mathrm{Aut}\, R \to \mathrm{Out}\, R$ is the projection, then by Theorem 1.6 the Q-kernels β' and $\Pi(\tilde{\alpha})\otimes\beta'$ are conjugate. Thus there exist unitaries $v_g, g\in G$ such that $(\mathrm{Ad}\, v_g\beta_g)_g$ is conjugate to $(\alpha_g\otimes\beta_g)_g$. Since α is conjugate to $\alpha\otimes\alpha$, Lemma 2.4 shows that β is outer conjugate to $\alpha\otimes\beta$, and hence $[\alpha]$ is a unit in Σ. The theorem is proved.

2.7 The above result extends to the following framework.

THEOREM. *Let* G *be a countable discrete amenable group and let* M *be a McDuff factor with separable predual. Two approximately inner actions* $\alpha,\beta\colon G \to \mathrm{Aut}\, M$ *with* $\alpha^{-1}(\mathrm{Int}\, M) = \alpha^{-1}(\mathrm{Ct}\, M) = \beta^{-1}(\mathrm{Int}\, M) = \beta^{-1}(\mathrm{Ct}\, M) = N$ *are outer conjugate if and only if* $\Lambda(\alpha)=\Lambda(\beta)$.

Proof. We again let $\Gamma = \Lambda(G,N)$ and let Δ be the set of outer conjugacy classes $[\alpha]$ of actions $\alpha\colon G \to \mathrm{Aut}\,\bar{M}$, with $\bar{M}$ isomorphic to M, $\alpha(G) \subset \overline{\mathrm{Int}\,\bar{M}}$ and $\alpha^{-1}(\mathrm{Int}\, M) = \alpha^{-1}(\mathrm{Ct}\, M) = N$. We let $\phi\colon\Delta\to\Gamma$ be the map $[\alpha]\to\Lambda(\alpha)$. For each $\xi\in\Gamma$, let $\alpha^{\xi}\colon G \to \mathrm{Aut}\, R$ be an action with $N(\alpha^{\xi})=N$ and $\Lambda(\alpha^{\xi})=\xi$. We let Γ act on Δ by

$$(\xi, [\alpha]) \to [\alpha^{\xi}\otimes\alpha] .$$

This map is well defined and we have

$$\phi([\alpha^{\xi}\otimes\alpha]) = \xi\phi([\alpha]) .$$

To apply Lemma 2.5(b) we have to show that Δ is a Γ-module and that $\phi^{-1}(1)$ has a single element. This last fact is established as in the proof of the preceding theorem, using Theorem 1.4 instead of its Corollary. In the same way we obtain that multiplication with the action of G on R coming from the free action of G/N on R preserves the class. The fact that for $\xi,\eta \in \Gamma$ and $[\beta] \in \Delta$,

$$[\alpha^{\xi} \otimes \alpha^{\eta} \otimes \beta] = [\alpha^{\xi\eta} \otimes \beta]$$

follows from the preceding theorem, since

$$\Lambda(\alpha^{\xi} \otimes \alpha^{\eta}) = \xi\eta = \Lambda(\alpha^{\xi\eta}) \quad .$$

The proof is thus finished.

2.8 For infinite factors we first need the following result.

LEMMA. *Let* $\alpha\colon G \to \operatorname{Aut} M$ *be an action of a discrete group of an infinite factor. Then* α *is outer conjugate to* $\mathrm{id}_F \otimes \alpha$ *where* F *is a type* I_∞ *factor.*

Proof. We let N be a type I_∞ subfactor of M. It is well known that $M = N \otimes (N' \cap M)$ and that there exists for each $g \in G$ a unitary $v_g \in M$ such that $\operatorname{Ad} v_g \alpha_g | N = \mathrm{id}_N$ (the proof of Lemma 8.4, step A, extends immediately to infinite dimensional subfactors).

Lemma 2.4 concludes the proof, since N is isomorphic to $N \otimes N$.

2.9 Let us now describe the classification of actions on the hyperfinite II_∞ factor $R_{0,1}$. There exists a homomorphism $\operatorname{mod}\colon \operatorname{Aut} R_{0,1} \to \mathbb{R}_+$ such that for $\theta \in \operatorname{Aut} R_{0,1}$ and τ a semifinite trace on $R_{0,1}$, $\tau \circ \theta = \operatorname{mod}(\theta)\tau$ [7]. It was shown by Connes [4] that $\operatorname{Ct} R_{0,1} = \operatorname{Int} R_{0,1}$ and $\overline{\operatorname{Int}}\, R_{0,1} = \ker \operatorname{mod}$.

For an action $\alpha\colon G \to \operatorname{Aut} R_{0,1}$ the homomorphism $\operatorname{mod}(\alpha)\colon G \to \mathbb{R}_+$ yields a conjugacy invariant; since inner automorphisms have module 1, $\operatorname{mod}(\alpha)$ is an outer conjugacy invariant.

THEOREM: *Let* G *be a countable discrete amenable group. Two actions* $\alpha,\beta\colon G \to \operatorname{Aut} R_{0,1}$ *are outer conjugate if and only if* Γ $(N(\alpha), \Lambda(\alpha), \operatorname{mod}(\alpha)) = (N(\beta), \Lambda(\beta), \operatorname{mod}(\beta))$.

Proof. We keep a normal subgroup N of G fixed and let Γ_0 be the group of all homomorphisms $\nu\colon G \to \mathbb{R}_+$ with $N \subseteq \ker \nu$. We let be the product of the groups $\Lambda(G,N)$ and Γ_0, and let Σ be the set of all outer conjugacy classes $[\alpha]$ of actions $\alpha: G \to \mathrm{Aut}\, M$ with M isomorphic to $R_{0,1}$, and with $N(\alpha) = N$. Since $R_{0,1} \simeq R_{0,1} \otimes R_{0,1}$ it is easy to see that Σ is a semigroup with multiplication given by the tensor product. The map $\psi: [\alpha] \to (\Lambda(\alpha), \mathrm{mod}(\alpha))$ yields a homomorphism of Σ into Γ. For $\xi \in \Lambda(G,N)$ let α^ξ be an action $G \to \mathrm{Aut}\, R$ with $\Lambda(\alpha^\xi) = \xi$. By results of Takesaki [42] there exists an action $\beta: \mathbb{R}_+ \to \mathrm{Aut}\, R_{0,1}$ with $\mathrm{mod}(\beta_t) = t$. For $\nu \in \Gamma_0$ we define an action $\beta^\nu: G \to \mathrm{Aut}\, R_{0,1}$ by $\beta^\nu_g = \beta_{\nu(g)}$. Then the action $\gamma = \alpha^\xi \otimes \beta^\nu$ of G on $R \otimes R_{0,1} \simeq R_{0,1}$ satisfies $N(\gamma) = N$, $\Lambda(\gamma) = \xi$ and $\mathrm{mod}(\gamma) = \nu$, hence ψ is surjective.

If $\psi([\alpha]) = 1$, then α is approximately inner and hence by Theorem 2.7, α is uniquely determined. Let $\alpha: G \to \mathrm{Aut}\, R_{0,1}$ with $\alpha \in \Sigma$ and let $\tilde{\alpha}: G \to \mathrm{Aut}\, R$ come from a free action of G/N on R. From Theorem 1.6 applied to the G/N-kernel induced by α, we obtain as in the proof of 2.6 the fact that α is outer conjugate to $\tilde{\alpha} \otimes \alpha$. On the other hand, by Lemma 2.8, α is outer conjugate to $\mathrm{id}_F \otimes \alpha$ where F is a type I_∞ factor. The class of the action $\tilde{\alpha} \otimes \mathrm{id}_F: G \to \mathrm{Aut}(R \otimes F) \simeq \mathrm{Aut}\, R_{0,1}$ thus acts as a unit in Σ. By Lemma 2.5(a), the theorem is proved.

2.10 The classification of G-kernels on factors can be done by the same methods, using Theorems 1.5 and 1.6 instead of their analogues 1.2 and 1.3 for actions. The key remark is that the Isomorphism Theorem 1.4 works for centrally free approximately inner G-kernels $\beta: G \to \mathrm{Out}\, M$ with trivial obstruction. By the definition of the obstruction, in this case there exists a cocycle crossed action (α, u) of G on M such that $(\beta_g) = (\Pi(\alpha_g))$ where $\Pi: \mathrm{Aut}\, M \to \mathrm{Out}\, M$ is the projection. Since G is amenable, u is a coboundary by Theorem 1.1, and one can suppose that α is an action; Theorem 1.4 can now be applied to conclude that the conjugacy class of β is uniquely determined. The existence of free G-kernels on R having arbitrary obstructions yields, in the same way as in 2.6, the following result.

THEOREM. *Let* G *be a countable discrete amenable group. Two free* G-*kernels* $\beta, \gamma: G \to \mathrm{Out}\, R$ *are conjugate if and only if* $\mathrm{Ob}(\beta) = \mathrm{Ob}(\gamma)$.

2.11 A result analogous to **2.7** is the following.

THEOREM. *Let* G *be as above and let* M *be a McDuff factor with separable predual. Two centrally free approximately inner* G-*kernels* $\beta, \gamma: G \to \mathrm{Out}\, M$ *are conjugate if and only if* $\mathrm{Ob}(\beta) = \mathrm{Ob}(\gamma)$.

2.12 Since inner automorphisms of $R_{0,1}$ have module 1, for a G-kernel $\beta: G \to \mathrm{Out}\, R_{0,1}$ the invariant $\mathrm{mod}(\beta): G \to \mathbb{R}_+$ can be defined. In the same way as in **2.10** one can prove the following result.

THEOREM. *Let* G *be a countable discrete amenable group. Two free G-kernels* $\beta, \gamma: G \to \mathrm{Out}\, R_{0,1}$ *are conjugate if and only if* $(\mathrm{Ob}(\beta), \mathrm{mod}(\beta)) = (\mathrm{Ob}(\gamma), \mathrm{mod}(\gamma))$.

Chapter 3: AMENABLE GROUPS

We associate to an amenable group G a paving system which is a system of finite sets that approximate the behavior of left G-spaces.

3.1 The group G which is dealt with in the sequel will be discrete, at most countable and nontrivial. G is called amenable if it has a left invariant mean, which is a positive linear map $m: \ell^\infty_{\mathbb{C}}(G) \to \mathbb{C}$, with $m(\mathbb{1}) = 1$ and $m \cdot \ell_g = m$ for $g \in G$, where ℓ_g is the left g translation on $\ell^\infty_{\mathbb{C}}(G)$; m is a "finitely additive finite Haar measure". For finite groups m is the Haar measure, but for infinite groups, such a mean, if it exists, is never unique. Abelian groups are amenable, since an invariant mean can be chosen by means of the Markov-Kakutani fixed point theorem. An ascending union of amenable groups is amenable, hence locally finite groups are amenable. Subgroups and quotient groups of amenable groups are amenable, and an extension of an amenable group with an amenable quotient is again amenable, hence solvable groups are amenable. The free group with two generators is not amenable. For a survey of amenability see [15]. If the group G is written as F/R with F a free group and R the relation subgroup, the amenability of G is connected to the "growth ratio" of R in F ([16]).

In what follows, for a set K we shall denote by $|K|$ its cardinality, and we shall write $K \subset\subset L$ if $K \subset L$ and $|K| < \infty$.

3.2 Let G be a group. If $F \subset\subset G$ and $\varepsilon > 0$ we say that a nonvoid subset S of G is (ε,F)-*(left) invariant* if it is finite and $|S \cap \bigcap_{g \in F} gS| > (1-\varepsilon)|S|$. The following intrinsic characterization of amenable groups was given in [12] by Følner. For a short proof, due to Namioka and Day, see [15].

THEOREM (Følner). *A group* G *is amenable if and only if it has arbitrarily (left) invariant subsets, i.e. if for any* $\varepsilon > 0$ *and* $F \subset\subset G$ *one can find an* (ε, F)*-invariant subset* S *of* G.

3.3 An impediment towards more elaborate constructions was the absence of a link between several approximately invariant subsets of G. A result in this direction was announced in [36]. We need that result in a slightly more precise form, which for convenience we prove in the sequel.

Let us consider, for instance, the case $G = \mathbb{Z}^2$. A large rectangle, which is approximately invariant to given translations has, moreover, the property that one can cover the group with translates of it, without gaps or overlappings. One cannot do the same thing with an arbitrarily shaped almost invariant subset, e.g. a "disc". Nevertheless it is possible to cover G, within a given accuracy ε, by using translates of a finite number N of "discs", provided each is very large with respect to the preceding one; moreover N depends only on ε.

We say that a system $(S_i)_{i \in I}$ of finite sets are ε-*disjoint*, $\varepsilon > 0$, if there are subsets $S_i' \subseteq S_i$, $i \in I$, such that $|S_i'| \geq (1-\varepsilon)|S_i|$, and $(S_i')_i$ are disjoint. We say that the system $K_1, \ldots, K_N$ of finite subsets of the group G ε-*pave* the finite subset S of G if there are subsets $L_1, \ldots, L_N$ of G, called *paving centers*, such that $\bigcup_i K_i L_i \subset S$, $(K_i L_i)_{i=1,\ldots,N}$ are disjoint and ε-*cover* S, i.e. $|S \setminus \bigcup_i K_i L_i| < \varepsilon |S|$, and moreover for each i, $(K_i \ell)_{\ell \in L_i}$ are ε-disjoint. If there are $\delta > 0$ and $K \subset\subset G$ such that $K_1, \ldots, K_N$ ε-pave any (δ, K)-invariant $S \subset G$ we call $K_1, \ldots, K_N$ an ε-*paving system of sets*.

THEOREM (Ornstein and Weiss). *Let* G *be an amenable group. For any* $\varepsilon > 0$ *there is* $N > 0$, *such that for any* $\gamma > 0$ *and* $F \subset\subset G$, *there is an* ε*-paving system* $K_1, \ldots, K_N$ *of subsets of* G, *with each* K_i *being* (γ, F)*-invariant.*

More precisely, for any $0 < \varepsilon < \frac{1}{2}$ let $N > \frac{4}{\varepsilon} \log \frac{1}{\varepsilon}$ and $\delta = (\frac{\varepsilon}{3})^N$; let $K\ , \ldots, K_N \_\ G$ be such that K_{n+1} is $(\delta |\bar{K}_n|^{-1}, \bar{K}_n^{-1})$-invariant, where $\bar{K}_n = \bigcup_{p \geq n} K_p$ and $n = 1, \ldots, N-1$. Then any $S \subseteq G$ which is $(\delta, \bigcup_n K_n)$-invariant is ε-paved by $K_1, \ldots, K_N$.

Remark. The essential fact is that N does not depend on the invariance degree (γ, F) imposed on the sets $(K_i)_i$.

The proof that follows is based on the ideas of Ornstein and Weiss.

The following lemma shows that if S is invariant enough with

with respect to K then it can swallow enough right translates of K; moreover, from the approximate invariance of S and K follows the approximate invariance of the remaining part, provided this part is not too small.

LEMMA. *Let* $0<\varepsilon<\frac{1}{2}$. *Suppose* $S\subset\subset G$ *is* $(\frac{1}{2},K)$*-invariant and let* $L\subset G$ *be maximal such that* $KL\subset S$ *and* $(K\ell)_{\ell\in L}$ *are* ε*-disjoint. Then* $|KL|\geq\frac{\varepsilon}{2}|S|$.

Suppose moreover that for some $\delta>0$ *and* $F\subset\subset G$, S *is* (δ,F)*-invariant and* K *is* $(\delta|F|^{-1},F^{-1})$*-invariant. If for* $\rho>0$, $|S\setminus KL|\geq\rho|S|$, *then* $S\setminus KL$ *is* $(3\rho^{-1}\delta,F)$*-invariant.*

Proof. Let $S'=S\cap\bigcap_{k\in K}k^{-1}S$; we have $|S'|\geq\frac{1}{2}S$. From the maximality of L it follows that for any $\ell\in S'$, $|K\ell\cap KL|\geq\varepsilon|K|$. In terms of characteristic functions this yields

$$\chi_{K^{-1}}*\chi_{KL}\geq\varepsilon|K|\chi_{S'}\ .$$

Integrating we get

$$|K|\,|KL|\geq\varepsilon|K|\,|S'|$$

hence

$$|KL|\geq\varepsilon|S'|\geq\frac{\varepsilon}{2}|S|$$

and the first part of the lemma is proved.

Suppose now that the supplementary assumptions are fulfilled, and let $S''=S\cap\bigcap_{k\in F}k^{-1}S$ and $K'=K\cap\bigcap_{k\in F}kK$; from the hypothesis $|S''|\geq(1-\delta)|S|$ and $|K'|\geq(1-\delta|F^{-1}|)|K|$. Let $S_1=S\setminus KL$ and $S_1'=S_1\cap\bigcap_{k\in F}k^{-1}S_1$. Then

$$S_1\setminus S_1'\subseteq(S\setminus S'')\cup(F^{-1}KL\setminus KL)\subseteq(S\setminus S'')\cup F^{-1}(K\setminus K')L\ .$$

So

$$|S_1\setminus S_1'|\leq|S\setminus S''|+|K\setminus K'|\,|F|\,|L|\leq\delta|S|+\delta|K|\,|L|\ .$$

From the ε-disjointness of $(K\ell)_{\ell\in L}$ it follows that

$$|KL|\geq(1-\varepsilon)|K|\,|L|$$

hence

$$|K|\,|L|\leq(1-\varepsilon)^{-1}|KL|\leq2|KL|\leq2|S|\ .$$

With the last hypothesis

$$|S_1\setminus S_1'|\leq3\delta|S|\leq3\delta\rho^{-1}|S_1|$$

and the lemma is proved.

Proof of the Theorem. Let $N > \frac{4}{\varepsilon}\log\frac{1}{\varepsilon}$, which implies $(1-\frac{\varepsilon}{2})^N < \varepsilon$. Let $\delta = (\frac{\varepsilon}{3})^N$ and $\delta_n = (\frac{\varepsilon}{3})^n$, $n = 1,\dots,N$. Let $S_N = S$ and for $n = N, N-1, \dots, 1$, supposing S_n is defined, let L_n be maximal such that $K_nL_n \subseteq S_n$ and $(K_n\ell)_{\ell\in L_n}$ are ε-disjoint; define $S_{n-1} = S_n \setminus K_nL_n$.

If for some $n \geq 1$, $|S_{n-1}| \leq \varepsilon|S_n|$, then we would be done since $|S \setminus \bigcup_p K_pL_p| = |S_0| \leq \varepsilon|S_N| = \varepsilon|S|$. We may therefore continue under the hypothesis $|S_{n-1}| > \varepsilon|S_n|$, $n = 1,\dots,N$ and we show inductively for $n = N, N-1, \dots, 1$ that

1) S_n is $(\delta_n, \bar{K}_n)$-invariant ;

2) $|S_{n-1}| \leq (1-\frac{\varepsilon}{2})|S_n|$.

By means of the lemma, (2) results from (1) since S_n is $(\frac{1}{2}, K_n)$-invariant. For n=N, (1) is a hypothesis. For $n < N$, since K_{n+1} is $(\delta_{n+1}|\bar{K}_{n+1}|^{-1}, \bar{K}_{n+1}^{-1})$-invariant and $|S_n| > \varepsilon|S_{n+1}|$, we infer that S_n is $(3\varepsilon^{-1}\delta_{n+1}\bar{K}_n) = (\delta_n, \bar{K}_n)$-invariant. An iteration of (2) shows that

$$|S_0| \leq (1-\frac{\varepsilon}{2})^N|S_N| \leq \varepsilon|S| .$$

The theorem is proved.

COROLLARY. *Suppose that a group* G *is infinite and* $K_1,\dots,K_N$ *are an* ε*-paving system of sets,* $\varepsilon > 0$. *There is a finite subset* K' *of* G, $K' \neq \emptyset$, *which may be chosen arbitrarily invariant such that there exist subsets* $L_1,\dots,L_n \subset\subset G$ *with*

1) $K' = \sum_{j=1}^{N} |K_i|\,|L_i|$;

2) *for any* $i \in \overline{1,N}$ *the sets* $\bar{K}_i' = \{g \in K' \mid \text{there are unique } (\bar{i},k,\ell) \in \coprod_{\bar{i}\in I} K_{\bar{i}} \times I_{\bar{i},j} \text{ with } g = k\ell, \text{ and for these } \bar{i} = i\}$ *satisfy* $|\bar{K}_i'| \geq (1-4\varepsilon)|K_i|\,|L_i|$.

Proof. Let $\varepsilon < \frac{1}{4}$. Suppose that $\tilde{K} \subset\subset G$ is ε-paved by $K_1,\dots,K_N$ with paving centers $\tilde{L}_1,\dots,\tilde{L}_N$. We may take $\tilde{K}$ arbitrarily invariant and $|\tilde{K}|$ arbitrarily large.

From the definition of ε-paving we easily infer

$$(1-\varepsilon)\sum_j |K_i|\,|L_i| \leq |\tilde{K}| \leq (1-\varepsilon)^{-1}\sum_j |K_i|\,|\tilde{L}_i| .$$

We can find $K' \subset G$ with $|K' \Delta \tilde{K}| < \sum_j |K_i|$ and for each $i \in \overline{1,N}$, $L_i \subset G$ with $|L_i \Delta \tilde{L}_i| < 2\varepsilon|\tilde{L}_i|$ such that $\sum_j |K_i|\,|L_i| = |K'|$. Since $|\tilde{K}|$ was as large as desired, there are no restrictions on the invariance degree of K'. One can still keep the assumption that $(K_iL_i)_i$ are mutually disjoint, and $(K_i\ell)$ for $\ell \in L_i$ are ε-disjoint; we obtain

3) $|K_iL_i \setminus K'| < 2\varepsilon |K_i|\,|L_i|$.

For $g \in G$ let $\psi_i(g) = |\{(k,\ell) \in K_i \times L_i \mid g = k\ell\}|$. The ε-disjointness condition yields

$$\{g \in K_iL_i \mid \psi_i(g) \leq 1\} > (1-2\varepsilon)|K_i|\,|L_i|$$

With (3) we infer

$$|K_i'| = |\{g \in K' \mid \psi_i(g) = 1\}| > (1-4\varepsilon)|K_i|\,|L_i| \ .$$

3.4 For an amenable group G, a repeated use of the Paving Theorem yields a sequence of "levels", each level consisting of a paving system of subsets of G which pave each of the sets appearing at the higher level. This structure contains all the information we need about the group G, and about the ways of approximating it with finite subsets. Therefore such a structure (fixed once and for all) will be the basis of all the constructions done further on. The proposition that follows is an immediate consequence of the Theorem and Corollary 3.3; the verifications are left to the reader.

PROPOSITION (Paving Structure). *Let* G *be an amenable group. Let* $\varepsilon_n > 0$ *and* $G_n \subset\subset G$ *be given, for* $n = 0,1,2,\ldots$. *Then there are* ε_n*-paving systems* $(K_i^n)_i$, $i \in I_n$, *with each* K_i^n *being* (ε_n, G_n)*-invariant and with* $(K_i^n)_i$ *mutually disjoint, and finite sets* $(L_{i,j}^n)_{i,j}$, $(i,j) \in I_n \times I_{n+1}$ *such that*

(1) $|K_j^{n+1}| = \sum_j |K_i^n|\,|L_{i,j}^n|$

and for any $(i,j) \in I_n \times I_{n+1}$, *the sets*

$$\bar{K}_{i,j}^{n+1} = \Big\{g \in K^{n+1} \mid \text{there are unique } (i,k,\ell) \in \coprod_{\bar{i}} K_{\bar{i}} \times L_{\bar{i},j} \text{ with } g = k\ell, \text{ and for these, } \bar{i} = i\Big\}$$

satisfy

(2) $|\bar{K}_{i,j}^{n+1}| \geq (1-\varepsilon_n)\,|K_i^n|\,|L_{i,j}^n|$.

Let $K^n = \coprod_i K_i^n$; since $(K_i^n)_i$ are supposed to be disjoint, we shall often identify K^n with $\bigcup_i K_i^n \subset G$.

For any n let $\bar{k}^n : \coprod_j \coprod_i K_i^n \times L_{i,j}^n \to \coprod_j K_j^{n+1} = K^{n+1}$ be a bijection such that for any $j \in I_{n+1}$, $\bar{k}^n\Big(\coprod_i K_i^n \times L_{i,j}^n\Big) = K_j^{n+1}$, and if $(i,k,\ell) \in \coprod_i K_i^n \times L_{i,j}^n$ with $k\ell \in \bar{K}_{k,j}^{n+1}$, then $\bar{k}^n(i,k,\ell) = k\ell$.

For any $g \in G$ and $n \in N$ let us choose "approximately left translations" with g in K^n, i.e. bijections $\ell_g^n : K^n \to K^n$ with $\ell_g^n(K_i^n) = K_i^n$, $i \in I_n$ such that if $k \in K_i^n$ with $gk \in K_i^n$, then $\ell_g^n(k) = gk$.

We call $\mathcal{K} = (\varepsilon_n, G_n, (K_i^n)_i, (L_{i,j}^n)_{i,j}, \bar{k}^n, (\ell_g^n)_g)_n$ a Paving Structure for G; the notation that appears in the statement of the proposition will frequently be used in the rest of the paper.

COROLLARY. *By the conditions of the proposition, for any* $g \in G_n$ *and* $(i,j) \in I_n \times I_{n+1}$,

$$(3) \quad |\{(k,\ell) \in K_i^n \times L_{i,j}^n \mid \bar{k}^n(\ell_g^n(k),\ell) \neq \ell_g^{n+1}(\bar{k}^n(k,\ell))\}| \leq 3\varepsilon_n |K^n| |L_{i,j}^n|$$

that is, on most of the K^{n+1}, *for a given* g *and for* n *large enough the left* g *translation almost coincides with the left* g *translation on the plaques* K_i^n *product with the identity on the set of paving centers* $L_{i,j}^n$.

<u>Proof</u>. Let Δ be the set in the left member of (3). If $(k,\ell) \in K_i^n \times L_{i,j}^n$ are such that $gk \in K_i^n$, and $k\ell, gk\ell \in \bar{K}_{i,j}^{n+1}$ then $\ell_g^n(k) = gk$, $\bar{k}^n(gk,\ell) = gk\ell$, $\bar{k}^n(k\ell) = k\ell$ and $\ell_g^{n+1}(k\ell) = gk$ and so $(k,\ell) \notin \Delta$. We infer

$$\Delta \subseteq (K_i^n \, g^{-1} K_i^n) \times L_{i,j}^n \cup (\bar{k}^n)^{-1} (K_i^n L_{i,j}^n \quad \bar{K}_{i,j}^{n+1}) \cup \ldots$$

$$\ldots \cup (g^{-1} \times \ell)(\bar{k}^n)^{-1} \quad (K_i^n L_{i,j}^n \quad \bar{K}_{i,j}^{n+1}) \quad .$$

The fact that K_i^n is (ε_n, G_n)-invariant together with (2) yields

$$|\Delta| \leq \varepsilon_n |K_i^n| |L_{i,j}^n| + 2\varepsilon_n |K_i^n| |L_{i,j}^n| = 3\varepsilon_n |K_i^n| |L_{i,j}^n|$$

and the corollary is proved.

<u>Example</u>. Let us consider for instance the case $G = \mathbb{Z}^N$. Let $K^n = ([-3^n-1)/2, (3^n-1)/2])^N \subset \mathbb{Z}^N$ and let $L^n = \{-3^n, 0, 3^n\}^N$. Then $K^n L^n = K^{n+1}$, so K^n are rectangles paving K^{n+1} with paving centers L^n and $\bar{k}^n$ is simply the product. The invariance degree of each K^n increases as much as desired with n. The approximate left G translations ℓ_g^n on K^n can be taken, for instance, to be the translations modulo $(3^n\mathbb{Z})^N$. In the general case, since G may have no analogue of rectangles we have to use several K_i^n at each level and take $\bar{k}^n$ and let ℓ_g^n be bijections behaving well on most of the points.

3.5 For later use we are now going to make some assumptions on the elements of the Paving Structure, made possible by the freedom of choice that we had at each step of its construction.

For a finite group G for each n we let $I_n = \{1\}$ and take $K_1^n = G_n = G$, $L_{1,1}^n = \{1\}$. In what follows we assume that G is infinite.

For each n we choose $G_{n+1} \subset G$ after $(K_i^n)_i$ were chosen. We may thus assume that

$$\Big(\bigcup_{p<n} \bigcup_{i,j} L_{i,j}^p\Big) \cup \Big(\bigcup_{p\leqslant n} \bigcup_{i,j} K_i^p (K_i^p)^{-1} K_j^p (K_j^p)^{-1}\Big) \cup (G_n \cup \{1\}) \cup \Big(\bigcup_{p\leqslant n} \bigcup_i K_i^p\Big) \subset G_{n+1}$$

and also that

$$\bigcup_n G_n = G \quad .$$

Since for each $j \in I_{n+1}$, $|K_j^{n+1}| = \sum_i |K_i^n| \, |L_{i,j}^n|$ we may assume, by taking all $|K_i^n|$ large, that

$$\sum_{i,j} |L_{i,j}^n| \leqslant \tfrac{1}{2} \varepsilon_n |K_j^{n+1}| \quad .$$

Since we may also suppose that $|G_{n+1}| \leqslant \frac{1}{2} \varepsilon_n |K_j^{n+1}|$, we may assume that

$$|G_{n+1} \cup \bigcup_{i,j} L_{i,j}^n| \leqslant \varepsilon_n |K_j^{n+1}| \quad .$$

After the choice of $(K_j^{n+1})_j$ and of $(L_{i,j}^{n+1})_{i,j}$ we may, without interfering with the previous assumptions, replace them with $(K_j^{n+1} g_j)_j$, respectively $((g_j)^{-1} L_{i,j}^n)$, for some arbitrary elements $g_j \in G$. G being assumed infinite, we may use this device to assume that for each n, $(K_j^{n+1})_j$ are mutually disjoint and, moreover, that

$$\Big(\bigcup_j K_j^{n+1}\Big) \cap \Big(\bigcup_{p<n} \bigcup_{i,j} L_{i,j}^p\Big) = \emptyset \quad .$$

For each n, $\varepsilon_n > 0$ could be chosen arbitrarily small. We use this to simplify the constants appearing in the computations. To avoid a long and unhelpful list of assumptions of the form

$$\varepsilon_{n+1} < f(n, \varepsilon_0, \varepsilon_1, \ldots, \varepsilon_n)$$

with $f > 0$, we leave it to the reader to check the fact that for each n, a finite number of such assumptions are done in the rest of the paper.

A last problem: The final choice of the Paving Structure will be done in the next chapter, by possibly taking only a subsequence $(K_i^{n_p})_p$ of the levels. It is easy to see, due to the finite number of possible refinements of a finite number of levels, that assumptions can be made in such a manner that the assumptions appearing above are true for any such refinement.

Chapter 4: THE MODEL ACTION

For a given amenable group G we construct, based on the paving structure previously displayed, a faithful representation of G into the weak closure of an UHF-algebra, the representation being well provided with finite dimensional approximations. We obtain from it a model of free action of G on the hyperfinite II_1 factor.

For finite G, the model reduces essentially to the equivariant matrix units model of Jones ([23]) while for $G = \mathbb{Z}$ it is different from the one used by Connes in [4] and could be viewed as a noncommutative version of the odometer model used in ergodic theory.

4.1 Let G be an amenable group and let $\mathcal{K}$ be a paving structure of G, constructed as shown in the previous chapter, and fixed in all that follows. We define the *limit space* K^* of $\mathcal{K}$ to be the inductive limit of the system $(K^n)_{n\in\mathbb{N}}$ with maps $K^{n+1} \to K^n$ given by

$$K^{n+1} \longrightarrow \coprod_j \coprod_i K_i^n \times L_{i,j}^n \longrightarrow \coprod_i K_i^n = K^n$$

where the first map is the inverse of the bijection $\bar{k}^n$, and the second one is the natural projection. Thus the elements of K^* are fibers $(k_n)_n \in \prod_n K^n$ which satisfy for each $n \in \mathbb{N}$: $k_n \in K_{i_n}^n$ and $\bar{k}^n(k_{n+1}) = (k_n, \ell_n) \in K_{i_n}^n \times L_{i_n,i_{n+1}}^n$.

Let $p_n\colon K^* \to K^n$ be the canonical projection. With the inductive limit of the discrete topologies on each K^n, K^* becomes a compact space and the Borel algebra of Borel measurable subsets is generated by $\bigcup_n \mathcal{B}_n$, with $\mathcal{B}_n = \{p_n^{-1}(S) \mid S \subseteq K^n\}$. For $n \in N$, let Γ_0^n consist of those permutations γ^n of K^n which are direct sums of permutations of each K_i^n, $i \in I_n$. Any $\gamma^n \in \Gamma_0^n$ uniquely determines a $\gamma^{n+1} \in \Gamma_0^{n+1}$ by $\bar{k}^n(\gamma^{n+1}(k_{n+1})) = (i_n, \gamma^n(k_n), \ell_n)$ if $\bar{k}^n(k_{n+1}) = (i_n, k_n, \ell_n)$. In this way we obtain a homomorphism from Γ_0^n into the automorphism group of K^*; we denote by $\Gamma^n \subset \operatorname{Aut} K^*$ its range and let $\Gamma = \bigcup_n \Gamma^n$. Being an ascending union of finite groups, Γ is amenable.

4.2 We choose an extremal measure μ in the set of all Γ-invariant probability Borel measures on K^*; this set is nonempty by the amenability of Γ. Being extremal, μ is Γ-ergodic.

Let χ_i^n be the characteristic function of $p_n^{-1}(K_i^n) \subseteq K^*$, and let $\mu_i^n = \int \chi_i^n \, d\mu$.

For $\mathcal{B}_n$-measurable functions $f: K^* \to \mathbb{C}$, which take the value f_k on $p_n^{-1}(\{k\})$, $k \in K^n$, the operators

$$f \mapsto |\Gamma^n|^{-1} \sum_{\gamma \in \Gamma^n} f$$

$$f \mapsto \sum_i |K_i^n|^{-1} \sum_{k \in K_i^n} f_k \cdot \chi_i^n$$

are both conditional expectations on the Borel algebra generated by $(\chi_i^n)_{i \in I_n}$, as one can easily check; hence they are equal. Since μ is Γ-invariant, this shows that μ is well determined by $(\mu_i^n)_{n,i}$.

For $i \in I_n$ and $j \in I_{n+1}$, $\chi_i^n \chi_j^{n+1}$ is the characteristic function of $p_n^{-1}(K_i^n) \cap p_{n+1}^{-1}(\bar{k}^n(K_i^n \times L_{i,j}^n))$. From the equality of the conditional expectations for $\mathcal{B}_{n+1}$ applied to $f = \chi_i^n$ we infer

$$|\Gamma^{n+1}|^{-1} \sum_{\gamma \in \Gamma^{n+1}} \chi_i^n \cdot \gamma = \sum_j |K_j^{n+1}|^{-1} |K_i^n| |L_{i,j}^n| \chi_j^{n+1}$$

$$= \sum_j \lambda_{i,j}^n \mu_j^{n+1}$$

where $\lambda_{i,j}^n = |K_i^n| |L_{i,j}^n| |K_j^{n+1}|^{-1}$.

For $n < m$, $i_n \in I_n$, $i_m \in I_m$, let

$$\lambda_{i_n, i_m}^{n,m} = \sum_{i_{n+1}, \dots, i_{m+1}} \lambda_{i_n, i_{n+1}}^n \lambda_{i_{n+1}, i_{n+2}}^{n+1} \cdots \lambda_{i_{m-1}, i_m}^{m-1}$$

so that, for instance, $\lambda_{i,j}^{n,n+1} = \lambda_{i,j}^n$. In a way similar to the one in the case $m = n+1$, for any $m > n$ one infers

$$(1) \quad |\Gamma^m|^{-1} \sum_{\gamma \in \Gamma^m} \chi_i^n \cdot \gamma = \sum_{j \in I^m} \lambda_{i,j}^{n,m} \chi_j^m .$$

The subsets Γ^m of the amenable group Γ have arbitrarily large invariance degree when m grows. The Mean Ergodic Theorem in L^1-norm applied to the Γ-ergodic measure μ gives

$$\lim_{m \to \infty} |\Gamma^m|^{-1} \sum_{\gamma \in \Gamma^m} \chi_i^n \cdot \gamma = \int \chi_i^n \, d\mu = \mu_i^n .$$

Hence from (1),

$$\lim_{m \to \infty} \sum_{j \in I^m} |\lambda_{i,j}^{n,m} - \mu_i^n| \mu_j^m = 0$$

and so, for any $n \in \mathbb{N}$

$$(2) \qquad \sum_{i\in I^n} \sum_{j\in I^m} |\lambda^{n,m}_{i,j} - \mu^n_i|\mu^m_j < \varepsilon_n$$

for all large enough m.

The measure μ being chosen once and for all, we make a last assumption on the paving system $\mathcal{K}$. By refining its levels, i.e. replacing $(K^n)_n$ by $(K^{n_p})_{n_p}$ for some subsequence $(n_p)_p$ of $\mathbb{N}$, we may suppose that (2) holds for any n and $m \geq n+1$. This can be done without renouncing any of the conditions imposed on the Paving Structure, in view of our assumptions stated in **3.5**.

<u>Remark</u>. The above inequality states that the proportion $\lambda^n_{i,j}$ of right translates of K^n_i in K^{n+1}_j almost doesn't depend on j. This might be quite surprising since $\lambda^n_{i,j}$ is in fact arbitrary. What actually happens is that the ergodic measure μ and the level refinement "choose" a part of the system $(K^n_i)_{n,i}$ for which $\lambda^n_{i,j}$ is almost independent of j; on the rest of the diagram, μ^m_j being small, the contribution of the corresponding terms in the sum (2) is negligible.

Let us fix bijections $\bar{s}^n_{i,j}: T^n_{i,j} \times S^n_i \to S^{n+1}_j$.

We may also suppose that for each $j \in I_{n+1}$ there is a set M^n_j such that with $M^n = \coprod_{j\in I_{n+1}} M^n_j$,

$$|\bar{S}^{n+1}| = |\bar{S}^n|\,|M^n| \qquad \text{and} \qquad |M^n_j| = \bar{\mu}^{n+1}_j |M^n| .$$

We infer

$$\begin{aligned} |L^n_{i,j}|\,|T^n_{i,j}| &= \lambda^n_{i,j}\, |K^{n+1}_j|\, |K^n_i|^{-1} |S^{n+1}_j|\, |S^n_i|^{-1} \\ &= \lambda^n_{i,j}\, \bar{\mu}^{n+1}_j\, |\bar{S}^{n+1}|\,(\bar{\mu}^n_i)^{-1}\ |S^n|^{-1} \\ &= \lambda^n_{i,j}\, \bar{\mu}^{n+1}_j (\bar{\mu}^n_i)^{-1}\, |M^n| \quad . \end{aligned}$$

Hence

$$\Big|\, |L^n_{i,j}|\, |T^n_{i,j}| - |M^n_j| \,\Big| = (\bar{\mu}^n_i)^{-1}\, \bar{\mu}^{n+1}_j\, |\lambda^n_{i,j} - \bar{\mu}^n_i|\, |M^n| \quad .$$

It is possible to choose subsets $P^n_{i,j} \subseteq L^n_{i,j} \times T^n_{i,j}$ and $R^n_{i,j} \subseteq M^n_j$ such that

$$|P^n_{i,j}| = |R^n_{i,j}| = \min\{|L^n_{i,j}|\,|T^n_{i,j}|\,|M^n_j|\}$$

and a bijection $\bar{p}^n_{i,j}: R^n_{i,j} \to P^n_{i,j}$. We have

$$\coprod_{i,j} K^n_i \times S^n_i \times R^n_{i,j} \subseteq \coprod_i K^n_i \times S^n_i \times \Big(\coprod_j M^n_j\Big) = \bar{S}^n \times M^n$$

and

$$\coprod_{i,j} K_i^n \times P_{i,j}^n \times S_i^n \subseteq \coprod_{i,j} K_i^n \times L_{i,j}^n \times T_{i,j}^n \times S_i^n \xrightarrow{\sim} \coprod_j K_j^{n+1} \times S_j^{n+1} = S^{n+1}$$

where the last map is $\coprod_{i,j} \bar{k}^n \times \bar{s}_{i,j}^n$, $\bar{k}^n$ being defined in **2.5** and $\bar{s}_{i,j}^n$ above. As $|S^{n+1}| = |S^n||M^n|$ and $|P_{i,j}^n| = |R_{i,j}^n|$ there is a bijection

$$\pi^n\colon \bar{S}^n \times M^n = \Big(\coprod_i K_i^n \times S_i^n\Big) \times \Big(\coprod_j M_j^n\Big) \to \bar{S}^{n+1} = \coprod_j K_j^{n+1} \times S_j^{n+1}$$

satisfying for any $i \in I_n$, $j \in I_{n+1}$, $k \in K_i^n$, $s \in S_i^n$, $r \in R_{i,j}^n$.

$$\text{(2)}\quad \pi^n(K_i^n \times S_i^n \times R_{i,j}^n) = \bar{k}^n \times \bar{s}_{i,j}^n(K_i^n \times P_{i,j}^n \times S_i^n)\ ,$$

$$\pi^n(k,s,r) = (\bar{k}^n \times \bar{s}_{i,j}^n)(k, \bar{p}_{i,j}^n(r), s)\ .$$

The inequality (1) shows that the cardinality of the elements in the argument or range of π^n not appearing in the above equality is small, i.e.

$$\text{(3)}\quad \sum_{i,j} |K_i^n||S_i^n|\,(|M_j^n|-|R_{i,j}^n|) + \sum_{i,j} |K_i^n|\,(|L_{i,j}^n||T_{i,j}^n|-|P_{i,j}^n|)\,|S_i^n|$$

$$\leqslant \sum_{i,j} |K_i^n||S_i^n|\,(\bar{\mu}_i^n)^{-1}\,\bar{\mu}_j^{n+1}|\lambda_{i,j}^n - \bar{\mu}_i^n|\,|M^n|$$

$$= \sum_{i,j} |\bar{S}^n|\bar{\mu}_j^{n+1}\,|\lambda_{i,j}^n - \bar{\mu}_i^n|\,|M^n| \leqslant \varepsilon_n|\bar{S}^{n+1}|\ .$$

4.4 We use the sets constructed in the previous chapter to index the matrix units of an UHF-algebra. Let $\mathcal{E}^0$ be a finite dimensional factor of dimension $|\bar{S}^0| = 1$ and for $n \geqslant 0$ let $\mathcal{Q}^n$ be a factor of dimension $|M^n|$ and let $\mathcal{E}^{n+1} = \mathcal{E}^n \otimes \mathcal{Q}^n$. Let $\mathcal{E}$ be the finite factor obtained as weak closure of the UHF-algebra $\overline{\bigcup_n \mathcal{E}^n}$ on the GNS representation associated to its canonical trace.

Modulo obvious identifications we may suppose that $\mathcal{E}^n \subseteq \mathcal{E}^{n+1} \subseteq \mathcal{E}$. Since $\pi^n\colon \bar{S}^n \times M^n \to \bar{S}^{n+1}$, $n \in \mathbb{N}$, are bijections, we can choose systems (E_{s_1,s_2}^n), $s_1,s_2 \in S^n$, of matrix units in $\mathcal{E}^n$, $n \in \mathbb{N}$, which are connected via π^n, i.e. such that

$$E_{s_1,s_2}^n = \sum_m E_{\bar{s}_1,\bar{s}_2}^{n+1}$$

with $m \in M^n$, $\bar{s}_1 = \pi^n(s_1,m)$, $\bar{s}_2 = \pi^n(s_2,m)$.

For any $g \in G$ and $n \geqslant 1$, the "approximate left g-translation" $\ell_g^n\colon K^n \to K^n$ defined in **3.4** yields a unitary $U_g^n \in \mathcal{E}^n$, given by

$$U_g^n = \sum_i \sum_{(k,s)} E^n_{(k_1,s),(k,s)}$$

where $i \in I_n$, $(k,s) \in K_i^n \times S_i^n$ and $k_1 = \ell_g^n(k)$. One can view U_g^n as the image of g in an "approximate left regular representation" of G. This is justified by the following proposition, which is the goal of all the constructions done before.

PROPOSITION. *Let* τ *be the canonical trace on* $\mathcal{E}$ *and* $|\cdot|_\tau$ *the corresponding* L^1*-norm. Then the limits*

$$U_g = \lim_{n\to\infty} U_g^n , \qquad g \in G$$

exist in $|\cdot|_\tau$*-norm and yield a faithful unitary representation of* G *into* $\mathcal{E}$. *For any* $n \geq 1$ *and* $g \in G_n \subset\subset G$ *(see* **3.4***) we have*

(1) $|U_g^n - U_g|_\tau \leq 8\varepsilon_n$.

Proof. In view of the fact that $G_n \nearrow G$, it is enough to prove the following inequalities

(2) $|U_g^n - U_g^{n+1}|_\tau \leq 7\varepsilon_n$ for $g \in G_n$,

(3) $|U_g^n U_h^n - U_{gh}^n|_\tau \leq 2\varepsilon_n$, for $g,h \in G_n$ with $gh \in G_n$,

(4) $|\tau(U_g^n)| \leq \varepsilon_n$ for $g \in G_n$, $g \neq 1$.

Statement (1) in the proposition is easy to obtain from (2), since in view of **3.5** we have $7\varepsilon_{n+1} + 7\varepsilon_{n+2} + \ldots < \varepsilon_n$.

Let us prove (4). For $g \in G$,

$$\tau(U_g^n) = |S^n|^{-1} \sum_{i\in I} |S_i^n| \, |\{k \in K_i^n \mid \ell_g^n(k) = k\}| \quad .$$

If $g \in G_n$, $g \neq 1$ and $k \in K_i^n \cap g^{-1}K_i^n$, then $\ell_g^n(k) = gk \neq k$. Since K_i^n is (ε_n, G_n)-invariant,

$$\tau(U_g^n) \leq |S^n|^{-1} \sum_{i\in I_n} |S_i^n| \, \varepsilon_n \, |K_i^n| = \varepsilon_n \quad .$$

Let us now prove (3). Let $g,h,gh \in G_n$. If $k \in K_i^n$ with $hk, ghk \in K_i^n$ then $\ell_g^n \ell_h^n(k) = \ell_{gh}^n(k) = ghk$. So from the (ε_n, G_n)-invariance of K_i^n, it follows that

(5) $|\{k \in K_i^n \mid \ell_g^n \ell_h^n(k) \neq \ell_{gh}^n(k)\}| \leq \varepsilon_n |K_i^n|$.

We have

$$|U_g^n U_h^n - U_{gh}^n|_\tau = \sum_i \Big| \sum_{(k,s)} E^n_{(k_2,s),(k_1,s)} E^n_{(k_1,s),(k,s)} - E^n_{(k_3,s),(k,s)} \Big|_\tau$$

$$\leq \Big| \sum_i \sum_{(k,s)} E^n_{(k_2,s),(k,s)} - E^n_{(k_3,s),(k,s)} \Big|_\tau$$

where $i \in I_n$, $(k,s) \in K_i^n \times S_i^n$, $k_1 = \ell_h^n(k)$, $k_2 = \ell_g^n(k_1)$ and $k_3 = \ell_{gh}^n(k)$; moreover, in the last member we sum only for those k for which $k_2 \neq k_3$. Hence (5) yields

$$(6) \quad |U_g^n U_h^n - U_{gh}^n|_\tau \leq |S^n|^{-1} \sum_i 2\varepsilon_n |K_i^n|\,|S_i^n| = 2\varepsilon_n .$$

We shall now use the results in **4.3** to prove (2). Let $g \in G_n$. From the definitions,

$$U_g^n = \sum_i \sum_{k,s} E^n_{(k_1,s),(k,s)} = \sum_{i,j} \sum_{k,s,m} E^{n+1}_{\bar{s}_1,\bar{s}} = \Sigma_1 + \Sigma_2$$

where $i \in I_n$, $(k,s) \in K_i^n \times S_i^n$, $k_1 = \ell_g^n(k)$, $j \in I_{n+1}$, $m \in M_j^n$, $\bar{s} = \pi^n(k,s,m)$, $\bar{s}_1 = \pi^n(k_1,s,m)$; in Σ_1 appear those terms for which $m \; R_{i,j}^n$ and in Σ_2 those for which $m \in M_j^n \setminus R_{i,j}^n$. We infer

$$|\Sigma_2|_\tau \leq \sum_{i,j} |K_i^n|\,|S_i^n|\,(|M_j^n| - |R_{i,j}^n|)\,|\bar{S}^{n+1}|^{-1} .$$

From the assumption 4.3(2) on the bijection π^n we infer

$$\Sigma_1 = \sum_{i,j} \sum_{k,\ell,t,s} E^{n+1}_{(\bar{k}_1,\bar{s}),(\bar{k},\bar{s})}$$

where $(k,\ell,t,s) \in K_i^n \times L_{i,j}^n \times T_{i,j}^n \times S_i^n$ satisfy $(\ell,t) \in P_{i,j}^n$, and $\bar{k} = \bar{k}^n(k,\ell)$, $\bar{k}_1 = \bar{k}^n(\ell_g^n(k),\ell)$, $\bar{s} = s_{i,j}^n(t,s)$.

On the other hand,

$$U_g^{n+1} = \sum_{i,j} \sum_{k,\ell,t,s} E^{n+1}_{(\bar{k}_2,\bar{s}),(\bar{k},\bar{s})} = \Sigma_1' + \Sigma_2'$$

where $(k,\ell,t,s) \in K_i^n \times L_{i,j}^n \times T_{i,j}^n \times S_i^n$, and $\bar{k} = \bar{k}^n(k,\ell)$, $\bar{k}_2 = \ell_g^{n+1}(k^n(k,\ell))$, $\bar{s} = \bar{s}_{i,j}^n(t,s)$. In Σ_1' we sum for $(\ell,t) \in P_{i,j}^n$ and in Σ_2' for $(\ell,t) \in L_{i,j}^n \times T_{i,j}^n \setminus P_{i,j}^n$. Therefore

$$|\Sigma_2'|_\tau \leq \sum_{i,j} |K_i^n|\,(|L_{i,j}^n|\,|T_{i,j}^n| - |P_{i,j}^n|)\,|S_i^n|\,|S^{n+1}|^{-1}$$

and from 4.3(3) we infer

$$|\Sigma_2|_\tau + |\Sigma_2'|_\tau \leq \varepsilon_n |\bar{S}^{n+1}|\,|\bar{S}^{n+1}|^{-1} = \varepsilon_n .$$

With Corollary 3.4 we obtain

$$|\Sigma_1 - \Sigma_1'|_\tau \leq 2 \sum_{i,j} \left|\{(k,\ell) \in K_i^n \times K_{i,j}^n \,|\, \bar{k}^n(\ell_g^n(k),\ell) \neq \ell_g^{n+1}(\bar{k}^n(k,\ell))\}\right| \, |T_{i,j}^n| \, |S_i^n| \, |\bar{S}^{n+1}|^{-1}$$

$$\leq 2\cdot 3\varepsilon_n \sum_{i,j} |K_i^n| \, |L_{i,j}^n| \, |T_{i,j}^n| \, |S_i^n| \, |\bar{S}^{n+1}|^{-1} = 6\varepsilon_n \quad .$$

Finally,

$$|U^n - U^{n+1}|_\tau \leq |\Sigma_1 - \Sigma_1'|_\tau + |\Sigma_2|_\tau + |\Sigma_2'|_\tau \leq 6\varepsilon_n + \varepsilon_n = 7\varepsilon_n$$

and (2) is proved.

Let $\mathcal{A}^n$ be the maximal abelian subalgebra of $\mathcal{E}^n$ generated by $(E^n_{\bar{s},\bar{s}})$, $\bar{s} \in \bar{S}^n$. Then $\mathcal{A}^n \subset \mathcal{A}^{n+1}$ and if we let $\mathcal{A}$ denote the weak closure of $\bigcup_n \mathcal{A}^n$ in $\mathcal{E}$, then $\mathcal{A}$ is a maximal abelian subalgebra in $\mathcal{E}$. The following result is a consequence of the proof of the proposition.

COROLLARY. *For any* $n \geq 1$ *and* $g \in G_n$, *there exists a projection* $p_g^n \in \mathcal{A}$ *such that* $\tau(p_g^n) \leq 8\varepsilon_n$ *and* $(1 - p_g^n)U_g = (1 - p_g^n)U_g^n$.

<u>Proof</u>. Let $g \in G_n$ and consider the projection in $\mathcal{A}$:

$$q_g^n = \sum_{s \in \Delta} E^{n+1}_{\bar{s},\bar{s}} \quad ,$$

with $\Delta = \{\bar{s} \in \bar{S}^{n+1} \mid E^{n+1}_{\bar{s},\bar{s}} U_g^n \neq E^{n+1}_{\bar{s},\bar{s}} U_g^{n+1}\}$.

Then $(1 - q_g^n)U_g^n = (1 - q_g^n)U_g^{n+1}$ and a careful inspection of the proof of the proposition reveals that in view of (2) we have actually shown that

$$|\Delta| \leq 7\varepsilon_n |\bar{S}^{n+1}| \quad .$$

Hence $\tau(q_g^n) \leq 7\varepsilon_n$, and if we let $p_g^n = \bigvee_{k \geq h} q_g^k$ then

$$(1 - p_g^n)U_g = (1 - p_g^n)U_g^n \qquad g \in G_n$$

and

$$\tau(p_g^n) \leq \sum_{k \geq n} \tau(q_g^k) \leq \sum_{k \geq n} 7\varepsilon_k \leq 8\varepsilon_n \quad .$$

The corollary is proved.

<u>Remark</u>. Some words about the ideas that lie behind the proof. Let $\tilde{E}^{n,i}_{k_1,k_2} = \sum_{s \in S_i^n} E^n_{(k_1,s),(k_2,s)}$ for $i \in I_n$ and $k_1, k_2 \in K_i^n$. Let $(\tilde{F}^{n,i}_{k_1,k_2})_{i,k_1,k_2}$ be matrix units for an AF-algebra $\mathcal{B} = \overline{\bigcup_n \mathcal{B}_n}$ which has as Bratelli diagram the Paving Structure $(K_i^n)_{n,i}$ (actually the numbers $(|K_i^n|)_{n,i}$), and for which $|L_{i,j}^n|$ gives the multiplicity of the arrow

$K_i^n \to K_j^{n+1}$. Let h_n be the homomorphism $\mathcal{B}_n \to \mathcal{E}$ which maps $\tilde{F}^{n,i}_{k_1,k_2}$ onto $\tilde{E}^{n,i}_{k_1,k_2}$. Then $h_{n+1}|\mathcal{B}_n$ is approximately equal to h_n, with even better approximation as n grows. What we did in **4.3** was an almost "ergodic" embedding of this AF-algebra into the UHF-algebra $\mathcal{E}$, motivated by the fact that it is much easier to reconstruct UHF-algebras inside a given W*-algebra than AF-algebras.

The corollary shows that on $K_j^{n+1} \simeq \coprod_i K_i^n \times L_i^n$ we have $\ell_g^{n+1} \simeq \ell_g^n \times \mathrm{id}$, and so we obtain at the limit a representation of G in the weak closure of $\mathcal{A}$. If $|I_n| = 1$ for all n, then $\mathcal{B}$ is an UHF-algebra and taking all multiplicities $|S_i^n|$ to be 1, we are done. If the proportion of $K_i^n L_{i,j}^n$ in K_j^{n+1} does not depend on j, we can still take the same multiplicities for all K_i^n and again we are done. In the general case in **4.2**, the ergodic measure μ on the topological dynamical system (K^*,Γ) yields a tracial factorial state on $\mathcal{B}$ by the construction of Krieger, Strătilă and Voiculescu [41]. In this way we obtain a finite hyperfinite factor and the combinatorics in **4.3** can be viewed as an explicit form of the classical proof of Murray and von Neumann [31] that such a factor is generated by an UHF-algebra.

4.5 Let us recall some notation and results in this chapter which are needed further on in the paper.

We have started with a discrete countable amenable group G, for which a Paving Structure was introduced in **4.3**. For $n \in \mathbb{N}$, with (K_i^n), $i \in I_n$, the ε_n-paving subsets of G on the n-th level of the Paving Structure, we have constructed finite sets (S_i^n), $i \in I_n$, and have set $\bar{S}^n = \bigcup_i K_i^n \times S_i^n$. We have considered a factor $\mathcal{E}^n$ with a matrix units basis $E_{s,t}^n$ indexed by $\bar{S}^n$ and have constructed unitaries U_g^n in $\mathcal{E}^n$, associated to the approximate left g-translation $\ell_g^n: \bigcup_i K_i^n \longrightarrow \bigcup_i K_i^n$ in the Paving Structure. We have denoted by $\mathcal{A}^n$ the maximal abelian subalgebra of $\mathcal{E}^n$ generated by $(E_{s,s}^n)$. We call $((E_{s,t}^n),(U_g^n))$ the n-*th finite dimensional submodel.*

We have assumed that $\mathcal{E}^n \subset \mathcal{E}^{n+1}$ in such a way that $\mathcal{A}^n \subset \mathcal{A}^{n+1}$, $n \in \mathbb{N}$, and have let $\mathcal{E}$ be the weak closure with respect to the trace of $\bigcup_n \mathcal{E}^n$, and $\mathcal{A}$ be the "diagonal" maximal abelian subalgebra of $\mathcal{E}$ generated by $\bigcup_n \mathcal{A}_n$. Since $|\bar{S}^n| \to \infty$, $\mathcal{E}$ is a II_1 hyperfinite factor. For each $g \in G$, $U_g = \lim_{n\to\infty} U_g^n$ *-strongly was shown to exist and yield a faithful representation of G in $\mathcal{E}$. For each n, $\mathcal{E} = \mathcal{E}^n \otimes ((\mathcal{E}^n)' \cap \mathcal{E})$ and $(\mathcal{E}^n)' \cap \mathcal{E}$ is a II_1 hyperfinite subfactor of $\mathcal{E}$ on which Ad U_g acts almost trivially. We call $(\mathcal{E},(U_g))$ the *submodel* and (Ad U_g) the *submodel action.*

We let R be a countably infinite tensor product of copies of the submodel factor $\mathcal{E}$, taken with respect to the normalized trace, and for each $g \in G$, we let $\alpha_g^{(0)}$ be the corresponding tensor product of copies of the submodel action $\mathrm{Ad}\, U_g$. Then R is the hyperfinite II_1 factor and $(\alpha_g^{(0)})$ is an action $G \to \mathrm{Aut}\, R$ by which Lemma 1.3.8 of Connes [3] is free. We call R the *model* and $\alpha^{(0)}: G \to \mathrm{Aut}\, R$ the *model action*.

Chapter 5: ULTRAPRODUCT ALGEBRAS

We study specific properties of ultraproduct algebras and use the machinery developed thus far to study ultraproduct type automorphisms.

5.1 In what follows M will be a W^*-algebra with separable predual. We denote by $\mathcal{U}(M)$ its unitary group and by $\mathrm{Proj}\, M$ its projections; M^h will be the hermitean part of M and M_1 its unit ball. We choose once and for all a free ultrafilter ω on $\mathbb{N}$.

Let us consider the following C^*-subalgebras of $l^\infty(\mathbb{N},M)$: $\mathcal{M}$, consisting of the constant sequences; $\mathcal{M}_\omega$, the ω-centralizing sequences (i.e. sequences $(x^\nu)_\nu$ with $\lim_{\nu\to\omega} \|[x^\nu,\phi]\| = 0$ for any $\phi \in M_*$); $\mathcal{I}_\omega$, the sequences ω-converging $*$-strongly to 0; $\mathcal{M}^\omega$, the normalizing algebra of $\mathcal{I}_\omega$. Both $\mathcal{M}$ and $\mathcal{M}_\omega$ normalize $\mathcal{I}_\omega$, hence are C^*-subalgebras of $\mathcal{M}^\omega$.

Let ϕ be a normal faithful state of M. A sequence $(x^\nu)_\nu \in l^\infty(\mathbb{N},M)$ is in $\mathcal{M}^\omega$ iff for any $\varepsilon > 0$ there is a $\delta > 0$ and a neighborhood W of ω in $\mathbb{N}$ such that for $y \in M$ with $\|y\| \leq 1$ and $\|y\|_\phi^\# < \delta$ we have $\|x^\nu y\|_\phi^\# + \|yx^\nu\|_\phi^\# < \varepsilon$, $\nu \in W$.

We consider the quotient C^*-algebras $M^\omega = \mathcal{M}^\omega/\mathcal{I}_\omega$ and $M_\omega = \mathcal{M}_\omega/\mathcal{I}_\omega$ and identify M with $(\mathcal{M}+\mathcal{I}_\omega)/\mathcal{I}_\omega$. This way M and M_ω are C^*-subalgebras of M^ω and $M \cap M_\omega = Z(M)$. Any $\phi \in M_*$ gives a form ϕ^ω on M^ω by $\phi^\omega((x^\nu)_\nu) = \lim_{\nu\to\omega} \phi(x^\nu)$; its restriction to M_ω will be denoted by ϕ_ω. For simplicity of notation, we write $\|\cdot\|_\phi$ and $\|\cdot\|_\phi^\#$ for the norms $\|\cdot\|_{\phi^\omega}$ and $\|\cdot\|_{\phi^\omega}^\#$ on M^ω.

LEMMA. *Let* $\phi \in M_*^+$ *be faithful and* $y \in M^\omega$. *Then* $(M^\omega)_1^h$ *is complete in the topology given by the seminorm* $x \to \|x\|_{\phi^\omega} + \|xy\|_{\phi^\omega}$.

Proof. The above topology being metrizable, it is enough to prove sequential completeness. Let $(x_n)_n \subset (M^\omega)_1^h$ be a sequence such that

$$\|x_{n+1}^\nu - x_n^\nu\|_\phi + \|(x_{n+1}^\nu - x_n^\nu)y^\nu\|_\phi < 2^{-n}$$

Let $(x_n^\nu)_\nu$, $(y^\nu)_\nu$ be representing sequences for x_n and y, with all $x_n^\nu \in M_1^h$. For each n we can modify x_n^ν for ν outside a neighborhood of ω such that

$$\|x^\nu_{n+1} - x^\nu_n\|_\phi + \|(x^\nu_{n+1} - x^\nu_n)y^\nu\|_\phi < 2^{-n}$$

holds for all n and ν. Then, ϕ being faithful, for each fixed ν, $(x_n)_n$ is s*-fundamental hence s*-converges to some $x^\nu \in M^h_1$, and $(x^\nu_n y^\nu)_n$ s*-converges to $x^\nu y^\nu$; so for all n,

$$\|x^\nu_n - x^\nu\|_\phi + \|(x^\nu_n - x^\nu)y\|_\phi \leqslant 2^{-n+1}$$

and it remains to show that $(x^\nu)_\nu$ is ω-normalizing. But this is true, since for $(t^\nu)_\nu \in \mathcal{I}_\omega$ with all $t^\nu \in M^h_1$, when $\nu \to \omega$ we infer

$$\|x^\nu t^\nu\|_\phi \leqslant \|t^\nu\|_\phi \longrightarrow 0$$

$$\|t^\nu x^\nu\|_\phi \leqslant \|t^\nu x^\nu_n\|_\phi + \|x^\nu_n - x^\nu\|_\phi$$
$$\leqslant \|t^\nu x^\nu_n\|_\phi + 2^{-n+1} \longrightarrow 2^{-n+1}$$

for any n.

We are now in a position to prove the following extension of [1, Theorem 2.9].

PROPOSITION. M^ω *is a* W*-*algebra and* M *and* M_ω *are* W*-*subalgebras of it. For any faithful* ϕ *in* M_*, ϕ^ω *is in* $(M_\omega)_*$ *and is faithful.*

<u>Proof</u>. Let ϕ be a faithful normal state on M. Let $x \geqslant 0$ in $(M^\omega)^+$ be represented by $(x^\nu)_\nu \subset M^+$. If $\phi^\omega(x^*x) = 0$ then $(x^\nu)_\nu \in \mathcal{I}_\omega$ and $x = 0$, so ϕ^ω is faithful. By means of the GNS construction associated to the faithful ϕ^ω, we may suppose that M^ω is a C*-subalgebra of some $B(H)$ having a separating cyclic vector ξ. We show that $(M^\omega)^h_1$ is so-closed.

Let $(x_i)_i \subset (M^\omega)^h_1$ be a so-fundamental net; for any $y \in M^\omega$, x_i is fundamental in the topology of the lemma before, and so there is $x^y \in (M^\omega)^h_1$ with $x_i\xi \to x^y\xi$, $x_i y\xi \to x^y y\xi$. As ξ is separating, x^y does not depend on y, and so x_i converges on the dense subset $M^\omega\xi$ of H to some $x \in (M^\omega)^h_1$; therefore M^ω is a W*-algebra. Being $\|\cdot\|_\phi$ complete, M is so-closed in M^ω, and hence is a W*-subalgebra.

For any $x \in (M^\omega)^h$ we have

$$\|[x,\phi^\omega]\| \leqslant 2\|x\|_\phi$$

$$\|[x,y]\|_\phi \leqslant 2\|x\|_\phi \|y\| + \|xy\|_\phi + \|xy^*\|_\phi \qquad \text{for } y \in M .$$

The left members are thus so-continuous seminorms on $(M^\omega)^h$. Since they vanish precisely for $x \in (M_\omega)^h$ we have proved that M_ω is a W*-

subalgebra of M .

<u>Problem</u>. Is it always true that $M' \cap M^\omega = M_\omega$?

For $x \in M^\omega$ we can define $\tau^\omega(x) = \omega - \lim_{\nu\to\omega} x^\nu \in M$, where $(x^\nu)_\nu$ is a representing sequence for x. Its restriction τ_ω to M_ω is a faithful normal trace with values in $Z(M)$. For $\phi \in M_*$ the restriction ϕ_ω of ϕ^ω to M_ω depends only on the restriction of ϕ to $Z(M)$, since $\phi_\omega(x) = \phi(\tau_\omega(x))$, $x \in M$.

5.2 Further on we shall deal with certain automorphisms of M_ω and M^ω constructed from the automorphisms of M. Suppose we are given a sequence $(\alpha^\nu)_{\nu\in\mathbb{N}}$ of automorphisms of M such that $\beta = \lim_{\nu\to\omega} \alpha^\nu$ exists in the u-topology. This yields an automorphism of $l^\infty(\mathbb{N},M)$ sending $(x^\nu)_\nu$ into $(\alpha^\nu(x^\nu))_\nu$. Since

$$\|\alpha^\nu(x^\nu)\|_\phi^2 \leqslant \|\phi\cdot\alpha^\nu - \phi\cdot\beta\|\,\|x^\nu\|^2 + |(\phi\cdot\beta)(x^{\nu *}x^\nu)| ,$$

this automorphism leaves $\mathcal{I}_\omega$ invariant, and hence gives an automorphism α of M^ω. As

$$\begin{aligned}\|[\phi, \alpha^\nu(x^\nu)]\| &= \|[\phi\cdot\alpha^\nu, x^\nu]\| \\ &\leqslant \|[\phi\cdot\beta, x^\nu]\| + 2\|\phi\cdot\alpha^\nu - \phi\cdot\beta\|\,\|x^\nu\|\end{aligned}$$

α leaves M_ω invariant. We call such automorphisms of M^ω or M_ω *semiliftable*; if $\alpha^\nu = \beta$ for all ν we call the automorphism $\alpha = (\alpha^\nu)_\nu$ of M^ω, respectively M_ω, *liftable* and denote it by β^ω, respectively β_ω.

For a semiliftable $\alpha = (\alpha^\nu)_\nu \in \mathrm{Aut}\, M^\omega$, with $\beta = \lim_{\nu\to\omega} \alpha^\nu$, if $\psi \in M_*$ and $x = (x^\nu)_\nu \in M^\omega$, then

$$\psi(\tau^\omega(\alpha(x))) = \lim_{\nu\to\omega} \psi(\alpha^\nu(x^\nu)) = \lim_{\nu\to\omega} \psi(\beta(x^\nu)) = \psi(\beta(\tau^\omega(x)))$$

therefore $\tau^\omega\cdot\alpha = \beta\cdot\tau^\omega$; in particular, semiliftable automorphisms of M_ω fixing the center of M are τ_ω preserving.

Recall ([4]) that an automorphism θ of M is called *properly outer* if none of its restrictions under a nonzero invariant central projection of M is inner. We let CtM denote the *centrally trivial* automorphisms of M, i.e. those $\theta \in \mathrm{Aut}\, M$ with $\theta_\omega = \mathrm{id} \in \mathrm{Aut}\, M_\omega$, and call $\theta \in \mathrm{Aut}\, M$ *properly centrally nontrivial* if none of its restrictions under an invariant central projection of M is centrally trivial.

For a discrete group G, a map $\alpha: G \to \mathrm{Aut}\, M$ is called *free* (respectively *centrally free*) if all α_g for $g \neq 1$ are properly outer (respectively properly centrally nontrivial).

If $U \in \mathcal{U}(M_\omega)$, then $\mathrm{Ad}\, U \in \mathrm{Aut}\, M^\omega$ is semiliftable. A broader class of semiliftable automorphisms is obtained from the approximately inner automorphisms of M. Let $\beta \in \overline{\text{Int } M}$ and let $(U^{\nu})_{\nu}$ be a sequence of unitaries of M with $\lim_{\nu\to\infty} \text{Ad } U^{\nu} = \beta$. It is easy to see that $(U^{\nu})_{\nu}$ represents a unitary U in M^{ω}, and $\text{Ad } U \in \text{Aut } M^{\omega}$ is semiliftable. Moreover, $\beta = \text{Ad } U|M$, but, of course, Ad U is not uniquely determined by β.

In his paper [4], A. Connes establishes connections between the automorphism group of a factor, the richness of its centralizing algebra and its property of being McDuff. These properties are essential for the constructions that follow.

THEOREM (A. Connes). *Let* M *be a factor with separable predual. The following are equivalent:*

(1) M *is McDuff, i.e.* $M \simeq M \otimes R$, *with* R *the hyperfinite* II_1 *factor.*

(2) $\overline{\text{Int } M}/\text{Int } M$ *is not abelian.*

(3) $\overline{\text{Int } M} \not\subset \text{Ct } M$.

(4) M_{ω} *is not abelian.*

(5) M_{ω} *is type* II_1.

5.3 We formalize below some useful tricks in M^{ω} that come from techniques of von Neumann, Dixmier, McDuff, and Connes.

The idea of the first one is to reindex representing sequences of a part of M^{ω} fast enough to make another part of M^{ω} behave like constant sequences with respect to it.

LEMMA (Fast Reindexation Trick). *Let* M *be a* W*-*algebra with separable predual and* $\omega \in \beta\mathbb{N}\setminus\mathbb{N}$. *Let* N *and* F *be countably generated sub* W*-*algebras of* M^{ω}, *and* $\mathcal{B}$ *a countable family of liftable automorphisms, leaving* N *invariant.*

*There is a normal injective *-homomorphism* $\Phi: N \to M^{\omega}$ *with*

(1) Φ *is the identity on* $N \cap M$

(2) $\Phi(N \cap M_{\omega}) \subseteq F' \cap M_{\omega}$

(3) $\tau^{\omega}(a\Phi(x)) = \tau^{\omega}(a)\tau^{\omega}(x)$, $\quad x \in N,\ a \in F$

(4) $\beta^{\omega}(\Phi(x)) = \Phi(\beta^{\omega}(x))$, $\quad x \in N,\ \beta^{\omega} \in \mathcal{B}$.

<u>Proof.</u> We may suppose that $M \subseteq N \cap F$. For natural n, we take finite subsets $N_n \subseteq N_{n+1}$ of N with $\tilde{N} = \bigcup_n N_n$ a unital *-algebra over $\mathbb{Q} + i\mathbb{Q}$, s-dense in N and globally fixed by $\mathcal{B}$, such that $\tilde{N} \cap M$ is w-dense in M and $\tilde{N} \cap M_{\omega}$ is w-dense in $\tilde{N} \cap M_{\omega}$; finite subsets $F_n \subseteq F_{n+1}$ of F with $\tilde{F} = \bigcup_n F_n$ w-dense in F, $\tilde{F} \cap M$ w-dense in M and $\tilde{F} \cap M$

w-dense in $F \cap M_\omega$; finite subsets $M_n \subseteq M_{n+1}$ of M_* with union norm dense in M_*; finite subsets $B_n \subseteq B_{n+1}$ of $\mathcal{B}$ with union $\mathcal{B}$.

For each $x \in N$ we choose a representing sequence $(x^\nu)_\nu$ such that for any $\nu \in \mathbb{N}$, $\|x^\nu\| \leq \|x\|$, $(x^*)^\nu = (x^\nu)^*$, $(\lambda x)^\nu = \lambda x^\nu$ for $\lambda \in \mathbb{C}$, and $x^\nu = x$ for $x \in M$.

Let ϕ be a faithful normal state on M. For each $x \in M^\omega$ and $n \in \mathbb{N}$ find $\delta_n > \delta_{n+1}(x) > 0$ and a neighborhood $W_n(x) \supseteq W_{n+1}(x)$ of ω in $\mathbb{N}$ such that

$$\|y\|_\phi^\# \leq \delta_n(x) \implies \|x^\nu y\|_\phi^\# + \|yx^\nu\|_\phi^\# \leq 1/n \quad \text{for} \quad \nu \in W_n(x) .$$

For $n \geq 1$ choose $p(n) \in \mathbb{N}$ such that $p(n) \geq n$ and

(5) $p(n) \in W_n(x), \quad x \in N.$

(6) $\|x^{p(n)} y^{p(n)} - (xy)^{p(n)}\|_\phi^\# \leq 1/n , \quad x,y \in N_n.$

(7) $\|[x^{p(n)}, a^n]\|_\phi^\# \leq 1/n , \quad x \in N_n \cap M_\omega, \ a \in F_n .$

(8) $|\psi(a^n x^{p(n)}) - \psi(a^n \tau^\omega(x))| \leq 1/n , \quad x \in N_n, \ a \in F_n, \ \psi \in M_n.$

(9) $\|\beta(x^{p(n)}) - (\beta^\omega(x))^{p(n)}\|_\phi^\# \leq 1/n , \quad x \in N_n, \ \beta \in \operatorname{Aut} M$ with $\beta^\omega \in B_n.$

We define ϕ on $\tilde{N}$ for $x = (x^n)_n$, letting $\Phi(x)$ be represented by $(x^{p(n)})_n$. By (5), $\Phi(x) \in M_\omega$, and from (6) and (8), Φ is a τ and hence $\|\cdot\|_\phi$ preserving homomorphism, so it extends to a normal injective $*$-homomorphism of N into M. The statements of the lemma are now straightforward to obtain.

5.4 We can reindex sequences of a part of M slow enough to make them behave like constants with respect to another part of M and to a family of semiliftable automorphisms.

LEMMA (Slow Reindexation Trick). *Let* M *be a* W^**-algebra with separable predual and* $\omega \in \beta\mathbb{N}\setminus\mathbb{N}$. *Let* N *and* F *be countably generated sub* W^**-algebras of* M^ω *and* $\mathcal{A}$ *a countable family of semiliftable automorphisms of* M, *such that if* $(\alpha_\nu)_\nu \subset \mathcal{A}$ *and* $\beta = \lim_{\nu\to\omega} \alpha_\nu$, *then* $\beta \in \mathcal{A}$ *and such that* $\mathcal{A}$ *leaves* N *invariant.*

There is a normal injective $*$*-homomorphism* $\Phi: N \to M$ *satisfying*

(1) Φ *is the identity on* $N \cap M$.

(2) $\Phi(N \cap M_\omega) \subset M_\omega$.

(3) $\Phi(N) \subset (F \cap M_\omega)' \cap M^\omega$

(4) $\tau^\omega(a\Phi(x)) = \tau^\omega(a)\tau^\omega(x) , \quad x \in N, \ a \in F .$

(5) $\alpha(\Phi(x)) = \beta(\Phi(x)) = \Phi(\beta(x)) , \quad x \in N, \ \alpha = (\alpha_\nu)_\nu \in \mathcal{A} , \ \beta = \lim_{\nu\to\omega} \alpha_\nu .$

Proof. We may again suppose that $M \subseteq N \cap F$. Choose N_n, F_n, M_n, ϕ and the representing sequences for the elements of N as in the previous lemma. Moreover, take finite subsets $A_n \subseteq A_{n+1} \subseteq \mathcal{A}$ with union $\mathcal{A}$, and representing sequences $(\alpha^\nu)_\nu$ for any $\alpha \in \mathcal{A}$ with all $\alpha^\nu = \beta$ if $\alpha = \beta^\omega$ for some $\beta \in \operatorname{Aut} M$. Take in the same way as before $\delta_n(x)$ and $W_n(x)$ for $x \in N$, and choose for any natural n, $p(n) \in \mathbb{N}$ such that

$$p(n) \in W_n(x) \ , \qquad x \in N_n$$

$$\| x^{p(n)} y^{p(n)} - (xy)^{p(n)} \|_\phi^\# \leq 1/n \qquad x,y \in N_n$$

$$\| x^{p(n)}, a \|_\phi^\# \leq 1/n \quad , \quad a \in F_n \cap M \ , \quad x \in N_n \cap M_\omega$$

$$| \psi(\tau^\omega(a) x^{p(n)}) - \psi(\tau^\omega(a) \tau^\omega(x) | \leq 1/n \ , \quad x \in N_n, \quad a \in F_n, \quad \psi \in M_n$$

$$\| \beta(x^{p(n)}) - (\beta^\omega(x))^{p(n)} \|_\phi^\# \leq 1/n \ , \quad x \in N_n, \ \beta \in \operatorname{Aut} M \text{ with } \beta^\omega \in A^n .$$

There are neighborhoods $V_n \subseteq V_{n+1}$ of ω in $\mathbb{N}$ with $V_1 = \mathbb{N}$, $\bigcap_n V_n = \emptyset$ and such that for any $\nu \in V_n$,

$$\| [x^{p(n)}, a^\nu] \|_\phi^\# \leq 1/n \ , \qquad x \in N_n \ , \ a \in F_n \cap M_\omega$$

$$| \psi(a^\nu x^{p(n)}) - \psi(\tau^\omega(a) x^{p(n)}) | \leq 1/n \ , \qquad x \in N_n \ , \ a \in F_n, \ \psi \in M_n$$

$$\| \alpha^\nu (x^{p(n)}) - \beta(x^{p(n)}) \|_\phi^\# \leq 1/n \ , \qquad \alpha = (\alpha^m)_m \in A_n \text{ and } \beta = \lim_{m \to \omega} \alpha^m .$$

We define $k: \mathbb{N} \to \mathbb{N}$ by $k(\nu) = p(n)$ if $\nu \in V_n \setminus V_{n+1}$, and for $x \in \bigcup_n N_n$, we let $\Phi(x)$ be represented by $(x^{k(\nu)})_\nu$. The remaining part of the proof is similar to the one of the preceding lemma.

5.5 In M^ω we can put together parts of several representing sequences to obtain a new representing sequence.

LEMMA (Index Selection Trick). *Let M be a W^*-algebra with separable predual and $\omega \in \beta\mathbb{N}/\mathbb{N}$. Let $\mathcal{C}$ be a separable sub C^*-algebra of $l^\infty(\mathbb{N}, M^\omega)$ and $\mathcal{A}$ a countable set of semiliftable automorphisms of M^ω, which acting term by term on $\mathcal{C}$, leave it globally invariant. Then there is a C^*-homomorphism $\Psi: \mathcal{C} \to M^\omega$ such that for any* $\tilde{x} = (x_n)_n \in \mathcal{C}$

(1) $\tau^\omega(\Psi(\tilde{x})) = w - \lim_{n \to \omega} \tau^\omega(x_n)$

(2) $\Psi(\tilde{x}) = x$ if $x_n = x$ for all n

(3) $\Psi(\tilde{x}) \in M_\omega$ if $x_n \in M_\omega$ for all n

(4) $\Psi(\tilde{y}) = \alpha(\Psi(\tilde{x}))$ for $\alpha \in \mathcal{A}$ and $\tilde{y} = (\alpha(x_n))_n$.

Remark. From (1), if for some faithful $\phi \in M_*^+$, $\lim_{n\to\omega} \|x_n\|_\phi^\# = 0$, then $\Psi(\tilde{x}) = 0$.

Proof. Let $C_n \subset C_{n+1}$, $n \in \mathbb{N}$ be finite subsets of $\mathcal{C}$ with union $\tilde{\mathcal{C}} = \bigcup_n C_n$ a unital dense sub *-algebra of $\mathcal{C}$ over $\mathbb{Q}+i\mathbb{Q}$, kept globally invariant by $\mathcal{A}$, and such that $\tilde{\mathcal{C}} \cap l^\infty(\mathbb{N},M_\omega)$ is norm-dense in $\mathcal{C} \cap l^\infty(\mathbb{N},M_\omega)$. Let $A_n \subset A_{n+1}$ be finite subsets of $\mathcal{A}$ with union $\mathcal{A}$ and $M_n \subset M_{n+1} \subset M_*$ be finite sets with union norm dense in M_*. Let ϕ be a faithful normal state on M. Choose for each $\alpha \in \mathcal{A}$ a representing sequence $(\alpha^\nu)_\nu$. For all $x \in M^\omega$ take representing sequences $(x^\nu)_\nu$, real $\delta_n(x) > 0$ and neighborhoods $W_n(x)$ of ω in $\mathbb{N}$ as in the lemmas above.

For any $n \geq 1$ we choose $p(n) \in \mathbb{N}$, $p(n) \geq n$, such that

$$(5) \quad |\psi(\tau^\omega(x_{p(n)})) - \lim_{m\to\omega} \psi(\tau^\omega(x_m))| \leq 1/n \ , \quad x = (x_m)_m \in C_n, \quad \psi \in F_n.$$

Let V_n, $n \in \mathbb{N}$ be neighborhoods of ω in $\mathbb{N}$, $V_n \supseteq V_{n+1}$, $V_1 = \mathbb{N}$, $\bigcap_n V_n = \emptyset$ be such that for $n > 1$,

$$(6) \quad V_n \subseteq W_n(x_{p(n)}), \quad \tilde{x} = (x_m)_m \in C_n$$

$$(7) \quad |\psi(\tau^\omega(x_{p(n)}) - \lim_{m\to\omega} \psi(\tau^\omega(x_m))| \leq 1/n \ , \quad \tilde{x} = (x_m)_m \in C_n \ , \quad \psi \in M_n$$

$$(8) \quad \|x_{p(n)}^\nu \, y_{p(n)}^\nu - (xy)_{p(n)}\|_\phi^\# \leq 1/n \ , \quad \tilde{x} = (x_m)_m \ , \quad \tilde{y} = (y_m)_m \in C_n$$

$$(9) \quad \|\alpha^\nu(x_{p(n)}^\nu) - (\alpha(x))_{p(n)}^\nu\|_\phi^\# \leq 1/n, \quad \tilde{x} = (x_m)_m \in C_n$$

$$(10) \quad \|[x_{p(n)}^\nu, \psi]\| \leq 1/n, \quad \tilde{x} = (x_m)_m \in C_n \cap l^\infty(\mathbb{N},M_\omega) \ , \quad \psi \in M_n \ .$$

We let $k: \mathbb{N} \to \mathbb{N}$ be defined by $k(\nu) = p(n)$ for $\nu \in V_n \setminus V_{n+1}$. For $\tilde{x} = (x_n)_n \in \tilde{\mathcal{C}}$ let $\Psi(\tilde{x}) \in M^\omega$ be represented by the sequence $(x_{k(\nu)}^\nu)_\nu$. That $\Psi(\tilde{x})$ is indeed in M^ω is shown by (6). We have $\|\Psi(\tilde{x})\| \leq \|\tilde{x}\|$ for all $\tilde{x} \in \tilde{\mathcal{C}}$, and so we may extend Ψ to all of $\mathcal{C}$ by continuity. The lemma now follows easily.

5.6 In what follows, we often have to work in the relative commutant in M of some constructions already done. We therefore need the following property.

Definition. We call $\theta \in \mathrm{Aut}\, M$ *strongly outer* if the restriction of θ to the relative commutant of any countable θ-invariant subset of M_ω is properly outer. A discrete group action α of G on M_ω is *strongly free* if all α_g, $g \neq 1$, are strongly outer.

Problem. Is any properly outer semiliftable automorphism of M strongly outer?

Partial affirmative answers are given in the sequel, extending results of A. Connes.

LEMMA. *Let* M *be a* W^**-algebra with separable predual and* $\omega \in \beta\mathbb{N}/\mathbb{N}$. *Let* $\alpha = (\alpha^\nu)_\nu$ *be a semiliftable automorphism of* M_ω *and* $\beta = \lim_{\nu\to\omega} \alpha^\nu$. *If* β *is properly centrally nontrivial, then* α *is strongly outer.*

Proof. Suppose that the restriction of α to $S' \cap M_\omega$ is not properly outer for some countable α-invariant $S \subset M_\omega$, and thus there is a non-zero $a \in S' \cap M_\omega$ with

$$\alpha(y)a = ay \quad , \qquad y \in S' \cap M_\omega \quad .$$

Let p be the central support of $\tau_\omega(|a|^2)$ in M, and $q = \bigvee_{k\in\mathbb{Z}} \beta^k(p)$.

Since β is properly centrally nontrivial and $\beta(q) = q$, there is some $z \in M_\omega$ with $qz = z$ and $\beta_\omega(z) - z \neq 0$. But $q|\beta_\omega(z) - z|^2 \neq 0$, so there is $k \in Z$ with $\beta^k(p)|\beta_\omega(z) - z|^2 \neq 0$. Let $x = (\beta_\omega)^{-k}(z)$; then $p|\beta_\omega(x) - x| \neq 0$.

We now use the Slow Reindexation Trick. Let $\mathcal{A} = \{\alpha, \beta^\omega\} \in \mathrm{Aut}(M^\omega)$, N the smallest W^*-subalgebra of M_ω that $\mathcal{A}$ leaves invariant and which contains x, and let F be the sub W^*-algebra of M_ω generated by a, p, and the countable subset S. We send x into $y = \Phi(x) \in M_\omega$ such that $y \in S' \cap M_\omega$, $ya = ay$, $\alpha(y) = \beta_\omega(y)$ and

$$\tau_\omega(|a^*|^2 |\beta_\omega(y) - y|^2) = \tau_\omega(|a^*|^2)\tau_\omega(|\beta_\omega(x) - x|^2) \quad .$$

From our choice of x,

$$p\tau_\omega(|\beta_\omega(x) - x|^2) = \tau_\omega(p|\beta_\omega(x) - x|^2) \neq 0 \quad .$$

As p is the central support of $\tau_\omega(|a|^2) = \tau_\omega(|a^*|^2)$, we obtain

$$\tau_\omega(|(\beta_\omega(y) - y)a|^2) = \tau_\omega(|a^*|^2 |\beta_\omega(y) - y|^2)$$

$$= \tau_\omega(|a^*|^2)\tau_\omega(|\beta_\omega(x) - x|^2) \neq 0 \ .$$

Hence $(\beta_\omega(y) - y)a \neq 0$, in contradiction with the fact that

$$(\beta_\omega(y) - y)a = \alpha(y)a - ya = \alpha(y)a - ay = 0 \quad .$$

5.7 Another case in which a semiliftable automorphism of M_ω is strongly outer is treated in the following lemma.

LEMMA. *Suppose* M *is a factor and let* $\alpha = (\alpha^\nu)_\nu$ *be a semiliftable automorphism of* M_ω, *such that* α^ν *is properly centrally nontrivial for all* ν. *Then* α *is strongly outer.*

Proof. Since M is a factor, τ_ω takes scalar values; let $\tau = \tau_\omega$ and let $|x|_\tau = \tau(|x|)$ for $x \in M_\omega$.

Claim. Let $\beta \in \mathrm{Aut}\, M$ be properly outer and let $q \in \mathrm{Proj}\, M_\omega$ be maximal such that $\tau(q\beta(q)) \leqslant \frac{1}{4}\tau(q)$. Then $q \vee \beta(q) \vee \beta^{-1}(q) = 1$.

Indeed, if not, by [4, Theorem 1.2.1] (or, alternatively, by the same reasoning as in the proof of Lemma 6.3 below) we get a projection $q' \neq 0$ with $q' \leqslant 1 - (q \vee \beta(q) \vee \beta^{-1}(q))$ and $\tau(q'\beta(q')) \leqslant \frac{1}{4}\tau(q')$. But then $(q' \vee \beta(q'))(q \vee \beta(q)) = 0$ and thus the maximality of q is contradicted by replacing it with $q + q'$. The claim is thus proved, and from it we infer $\tau(q) = \tau(\beta(q)) \geqslant 1/3$ and so $\tau((\beta(q) - q)^2) = 2\tau(q) - 2\tau(q\beta(q)) \geqslant 2 \cdot 1/3 \cdot (1 - 1/4) = 1/2$.

To prove the lemma let $S = (s_n)_n$ be a countable α-invariant $*$-subset of M_ω and suppose that $\alpha | S' \cap M_\omega$ is not properly outer, that is, there exists $a \in S' \cap M_\omega$, $a \neq 0$, such that

$$\alpha(x)a = ax \quad \text{for} \quad x \in S' \cap M_\omega \ .$$

Let ψ be a normal faithful state on M and let $(\psi_n)_n$ be a total subset of M_*. Let $(a^\nu)_\nu$ and $(s_n^\nu)_\nu$ be representing sequences for a and s_n respectively; $n = 1, 2, \ldots$. Let us keep $\nu \in \mathbb{N}$ fixed. By means of the preceding lemma the hypothesis yields that $\beta = (\alpha^\nu)_\omega \in \mathrm{Aut}\, M_\omega$ is properly outer, and thus by the Claim there exists a projection $q \in M_\omega$ with $\tau((\beta(q) - q)^2) \geqslant 1/2$. We remark that in the algebra M^ω we have

$$\tau^\omega(|\beta(q)a^\nu - a^\nu q|^2) = \tau^\omega(|(\beta(q) - q)a^\nu|^2) = \tau^\omega(|a^\nu|^2)\tau((\beta(q) - q))^2$$
$$\geqslant 1/2\, \tau^\omega(|a^\nu|^2) \ .$$

Hence we can pick out of a representing sequence for q an element $q^\nu \in M$ such that $\|q^\nu\| \leqslant 1$ and

$$\|\alpha^\nu(q^\nu)a^\nu - a^\nu q^\nu\|_\phi \geqslant 1/2 \|a^\nu\|_\phi$$

$$\|[q^\nu, \psi_k]\| \leqslant \frac{1}{\nu} \ , \qquad k = 1, \ldots, \nu$$

$$\|[q \ , s_k]\|^\# \leqslant \frac{1}{\nu} \ , \qquad k, \mu = 1, \ldots, \nu \ .$$

Then the sequence $(q^\nu)_\nu$ represents an element $\bar{q} \in S' \cap M_\omega$ satisfying

$$\|\alpha(\bar{q})a - a\bar{q}\|^2 \geqslant 1/2 \|a\|_\tau^2 \neq 0$$

and the contradiction thus obtained shows that α is strongly outer.

5.8 The following result appears, with a slightly different proof, in [13, Lemma B.5].

LEMMA. *Let* M *be a factor and let* $E \subset M$ *be a finite dimensional subfactor,* $1 \in E$. *Let* $\omega \in \beta\mathbb{N}\backslash\mathbb{N}$. *Then the inclusion* $E' \cap M \to M$ *induces an isomorphism* $(E' \cap M)_\omega \to M_\omega$.

COROLLARY.

(1) *If* M *is McDuff then* $E' \cap M$ *is McDuff.*

(2) *If* $\theta \in \mathrm{Aut}\, M \backslash \mathrm{Ct} M$ *and* $\theta(E) = E$, *then* $(\theta | E' \cap M) \in \mathrm{Aut}(E' \cap M) \backslash \mathrm{Ct}(E' \cap M)$.

Proof. Let $(e_{i,j})$, $i,j \in I$, be a system of matrix units generating E. For any $y \in M$, $\|y\| \leq 1$, we have

$$y = \sum_{i,j} e_{i,j} y_{i,j}$$

with $y_{i,j} = \sum_k e_{k,i} y e_{j,k} \in E' \cap M$; $\|y_{i,j}\| \leq 1$.

If $\phi \in M_*$ and $x \in E' \cap M$, then

$$[\phi, x](y) = \sum_{i,j} e_{i,j} [\phi, x](y_{i,j})$$

hence

$$\| [\phi, x] \| \leq |I|^2 \| [(\phi | E' \cap M), x] \|$$

and thus the inclusion $E' \cap M \to M$ induces an inclusion $(E' \cap M)_\omega \to M_\omega$.

Let $P\colon M \cap E' \to M$ be the conditional expectation

$$x \longrightarrow P(x) = |I|^{-1} \sum_{i,j} e_{i,j} x e_{j,i} \quad , \qquad x \in M .$$

If $x \in M$, then

$$P(x) - x = |I|^{-1} \sum_{i,j} e_{i,j} [x, e_{j,i}] \quad .$$

Hence, if $(x^\nu)_\nu \in M_\omega$, then $\lim_{\nu \to \omega} (P(x^\nu) - x^\nu) = 0$ *-strongly and so $(P(x^\nu))_\nu \sim (x^\nu)_\nu$. Thus P induces a map $M_\omega \to (E' \cap M)_\omega$ that is inverse to the one induced by the inclusion.

The lemma is proved.

Chapter 6: THE ROHLIN THEOREM

In this chapter we prove a Rohlin type theorem for a discrete amenable group G acting centrally freely on a von Neumann algebra. As a consequence, we show that if H is a normal subgroup of G, the Rohlin theorem holds for the action of the quotient G/H on the almost fixed points for H.

6.1 Some of the basic tools in the modern developments of the ergodic theory in both measure spaces and von Neumann algebras are the various extensions of the Rohlin Tower Theorem. The one proved in the sequel essentially states that for a free enough action of a discrete amenable group G on a von Neumann algebra M, one can find a partition of the unity in projections indexed by finite subsets $(K_i)_i$ of G, such that G acts on it approximately the same way in which it acts on $\ell^\infty(\bigcup_i K_i)$ by means of the left regular action. The equivariant partition of unity thus obtained is the starting point of most of the constructive proofs that follow.

This theorem extends, on the one hand, Ornstein and Weiss's Rohlin Theorem for discrete amenable groups acting freely on a measure space ([36]) and, on the other hand, the Rohlin Theorem of Connes for single automorphisms of von Neumann algebras ([4]). For (not necessarily centrally-) free actions the theorem of Connes was extended in [33] to abelian groups, but for amenable groups this problem is still open.

If ϕ is a trace on the von Neumann algebra we let $|x|_\phi = \phi(|x|)$, $x \in M$. For the sake of simplicity, we write $|x|_\phi$ for $|x|_{\phi_\omega}$ if $x \in M_\omega$. Recall that a crossed action of G on M is a map $\alpha: G \to \operatorname{Aut} M$ with $\alpha_1 = 1$ and $\alpha_g \alpha_h \alpha_{gh}^{-1} \in \operatorname{Int} M$, $g,h \in G$.

THEOREM (Nonabelian Rohlin Theorem). *Let* G *be a discrete countable amenable group, let* M *be a von Neumann algebra with separable predual, and let* $\omega \in \beta\mathbb{N}/\mathbb{N}$. *Let* $\alpha: G \to \operatorname{Aut} M_\omega$ *be a crossed action which is semiliftable and strongly free. Let* ϕ *be a faithful normal state on* M *such that* $\alpha|Z(M)$ *leaves* $\phi|Z(M)$ *invariant. Let* $\varepsilon > 0$ *and let* $K_1, \ldots, K_N$ *be an* ε*-paving family of subsets of* G. *Then there is a partition of unity* $(E_{i,k})_{i=1,\ldots,N;\, k \in K_i}$ *in* M_ω *such that*

(1) $$\sum_{i=1}^{N} |K_i|^{-1} \sum_{k,\ell \in K_i} |\alpha_{k\ell^{-1}}(E_{i,\ell}) - E_{i,k}|_\phi \leq 5\varepsilon^{1/2};$$

(2) $[E_{i,k}, \alpha_g(E_{j,\ell})] = 0$ *for all* g,i,j,k,ℓ;

(3) $\alpha_g\alpha_h(E_{i,k}) = \alpha_{gh}(E_{i,k})$ *for all* g,h,i,k.

Moreover, $(E_{i,k})_{i,k}$ *can be chosen in the relative commutant in* M_ω

of any given countable subset of M_ω.

The estimate (1) above is an average estimate. Below we give other types of estimates that can be derived from it.

COROLLARY. *In the conditions of the theorem we have for any* $g \in G$

$$(4) \quad \sum_i \sum_k |\alpha_g(E_{i,k}) - E_{i,gk}|_\phi \leq 10\varepsilon^{\frac{1}{2}}, \quad i=1,\dots,N; \quad k \in K_i \cap g^{-1}K_i .$$

For any $\delta > 0$ *and any subsets* $A_k \subset K_i$ *with* $|A_i| \leq \delta|K_i|$, $i=1,\dots,N$, *we have*

$$(5) \quad \sum_i \sum_k |E_{i,k}|_\phi \leq \delta + 5\varepsilon^{\frac{1}{2}}, \quad i = 1,\dots,N; \quad k \in A_i.$$

<u>Proof</u>. For any $i=1,\dots,N$, $k \in K_i \cap g^{-1}K_i$ and $\ell \in K_i$,

$$|\alpha_g(E_{i,k}) - E_{i,gk}|_\phi \leq |\alpha_g(\alpha_{k\ell^{-1}}(E_{i,\ell}) - E_{i,k})|_\phi + |\alpha_{gk\ell^{-1}}(E_{i,\ell}) - E_{i,gk}|_\phi .$$

Summing for all k,ℓ as above, we infer

$$|K_i| \sum_k |\alpha_g(E_{i,k}) - E_{i,gk}|_\phi \leq 2 \sum_{\ell,m} |\alpha_{m^{-1}}(E_{i,m}) - E_{i,\ell}|_\phi$$

where $k \in K_i \cap g^{-1}K_i$ and $\ell,m \in K_i$. Hence (4) follows from (1).

Let us now prove (5). For any $i=1,\dots,N$, $m \in A_i$ and $k \in K_i$,

$$|E_{i,m}|_\phi \leq |E_{k,k}|_\phi + |\alpha_{mk^{-1}}(E_{i,k}) - E_{i,m}|_\phi .$$

Summing for all such m,k we get

$$|K_i| \sum_m |E_{i,m}|_\phi \leq |A_i| \sum_k |E_{i,k}|_\phi + \sum_{k,\ell} |\alpha_{k\ell^{-1}}(E_{i,\ell}) - E_{i,k}|_\phi$$

and thus

$$\sum_m |E_{i,m}|_\phi \leq \delta \sum_k |E_{i,k}|_\phi + |K_i|^{-1} \sum_{k,\ell} |\alpha_{k\ell^{-1}}(E_{i,\ell}) - E_{i,k}|_\phi$$

where $m \in A_i$, $k,\ell \in K_i$. Thus (5) is obtained from (1).

Here are some circumstances under which the hypothesis of the theorem is fulfilled:

If the algebra M is a factor, no assumption on the state ϕ is needed, since ϕ_ω is the canonical trace on M_ω, and is preserved by semiliftable automorphisms.

In the case when $\alpha: G \to \operatorname{Aut} M$ is induced by a centrally free crossed action $G \to \operatorname{Aut} M$, then by Lemma 5.7, α is a strongly free action; for instance if M is the hyperfinite II_1 of II_∞ factor, then any free action $G \to \mathrm{Aut}\, M$ is centrally free.

For an abelian algebra $M = L(X, \mathcal{B}, \mu)$, with μ a probability measure, if α is induced by a measure preserving free action of G on X, then α_ω is strongly free and one gets the Ornstein and Weiss theorem.

6.2 The proof of the theorem consists of two parts. In the first part we use a global geometric approach based on a lemma of Sorin Popa to obtain a basis for some (possibly small) Rohlin tower in M_ω. In the second part of the proof we put together such towers in order to get a Rohlin tower filling almost all the space. A difference between this part of the proof and the ones in [4] and [33] is that each time a new tower is added, one destroys a part of the old one, taking care to make the procedure convergent.

Let us first state the following result ([37, Lemma 1.3]) of Sorin Popa.

Let A be a finite von Neumann algebra with a finite normal faithful trace τ, and let B be a von Neumann subalgebra of A. Then there is a unique τ-preserving conditional expectation P_B of A onto B. One calls $x \in A$ orthogonal on B if $P_B(x) = 0$ (or equivalently if $\tau(xy) = 0$ for any $y \in B$).

LEMMA (S. Popa). *Let* A *be a finite von Neumann algebra,* τ *a normal faithful trace on it, and* B *a von Neumann subalgebra of* A. *Suppose that the relative commutant condition* $B' \cap A \subseteq B$ *holds. If* $\varepsilon > 0$ *and* $x_1, \ldots, x_m \in A$ *are orthogonal to* B, *then there exists a partition of unity* $(e_j)_{j=1,\ldots,n}$ *in* B *such that*

$$(1) \qquad \Big\| \sum_{j=1}^{n} e_j x_i e_j \Big\|_\tau \leq \varepsilon \| x_i \|_\tau \qquad \textit{for} \quad i = 1, \ldots, m .$$

Let us briefly sketch his proof, since in our context it will yield a geometrical insight into the structure of discrete crossed products.

One begins by proving an elementary Hilbert space lemma, asserting that if (U_g) is a unitary representation of a discrete group Γ on the Hilbert space H, which has no nontrivial fixed points in H, then for any $\xi \in H$ and $\delta > 0$ there exists $g \in \Gamma$ such that $U_g \xi$ is δ-orthogonal to ξ, i.e. such that $\|U_g \xi - \xi\| \geq (\sqrt{2} - \delta)\, \|\xi\|$. If not, one shows that for $\xi \neq 0$ the minimal norm point in $\overline{\mathrm{co}}^{w}\{U_g \xi \mid g \in \Gamma\}$ is nonzero and is fixed by (U_g).

Let $\rho : A \to \mathcal{B}(L^2(A, \tau))$ be the GNS representation and let U be the representation of the unitary group of B induced by ρ on the space $H = L^2(A,\tau)\ominus L^2(B,\tau)$. The absence of nontrivial fixed points for U follows from the relative commutant condition $B'\cap A \subseteq B$. The Hilbert space lemma yields for any $x\in A$ orthogonal on B (viewed as a vector in H) a unitary $u\in B$ with $\|uxu^* - x\|_\tau \geq \|x\|_\tau$. Spectral projections of u yield a first version of the looked for $e_1,\dots,e_n \in B$, with $n=1$ and $\varepsilon = \sqrt{3/4}$ in (1), and an inductive refinement of the procedure yields the results in the lemma.

Let us now consider the case when B is a finite von Neumann algebra with normalized trace τ, and (α_g) is a free τ-preserving action of a discrete group G on B.

Let A be the crossed product $B\times_\alpha G$ and let $\lambda_g \in A$ denote the unitary corresponding to the left g translation in $L(G)$. We identify B with $B\lambda_1$ and extend τ to a trace on A letting, for $x\in A$, $x = \sum_g x_g\lambda_g$, with $x_g\in B$, $\tau(x)=\tau(x_1)$. Then $x\to x_1$ is a τ-preserving conditional expectation of A onto B, and all λ_g for $g\neq 1$ are orthogonal on B.

Let $a = \sum_g a_g\lambda_g \in B'\cap A$. Then for any $x\in B$ and $g\in G$ we have $a_g\alpha_g(x) = xa_g$. Since α was assumed free, $a_g = 0$ for $g\neq 1$, and hence $B'\cap A\subseteq B$. This yields the following

COROLLARY. *Let* B,τ *and* $\alpha: G\to \mathrm{Aut}\, B$ *be as above. Let* $\delta>0$ *and let* K *be a finite subset of* G, *with* $1\notin K$. *Then there exists a partition of unity* $(e_j)_{j=0,1,\dots,n}$ *in* B *such that* $|e_0| < \delta$ *and*

$$|e_j\alpha_g(e_j)|_\tau < \delta|e_j|_\tau \qquad j=1,\dots,n;\quad g\in K .$$

<u>Proof</u>. In view of the preceding discussion we may apply Popa's lemma to the B-orthogonal family $\{\lambda_g | g\in K\}$ to get a partition of unity $(f_i)_{i\in I}$ in B with

$$\sum_i \|f_i\lambda_g f_i\|_\tau \leq \varepsilon^2|K|^{-1} , \qquad g\in K .$$

Thus for $g\in K$ we have

$$\sum_i \|f_i\alpha_g(f_i)\|_\tau = \sum_i \tau(f_i\alpha_g(f_i)f_i)^{\frac12} = \sum_i \tau(f_i\lambda_g f_i\lambda_g^* f_i)^{\frac12}$$
$$= \sum_i \|f_i\lambda_g f_i\|_\tau \leq \varepsilon^2|K|^{-1}$$

and by the Cauchy-Schwartz inequality

$$\sum_i |f_i\alpha_g(f_i)|_\tau \leq \sum_i \|1\|_\tau\|f_i\alpha_g(f_i)\|_\tau \leq \varepsilon^2|K|^{-1} .$$

Let $I_0 = \{i\in I \,\big|\, |f_i\alpha_g(f_i)|_\tau \geq \varepsilon\tau(f_i)$ for some $g\in K\}$. We infer

$$\varepsilon \sum_{i\in I_0} \tau(f_i) \leqslant \sum_{g\in K} \sum_{i\in I_0} |f_i\alpha_g(f_i)|_\tau < |K|\varepsilon^2|K|^{-1} = \varepsilon^2$$

and so, if $e_0 = \sum_{i\in I_0} f_i$, then $\tau(e_0) < \varepsilon$.

For any $i \in I\setminus I_0$ we have

$$|f_i\alpha_g(f_i)|_\tau < \varepsilon|f_i|_\tau \quad , \qquad g\in K$$

and all that remains to be done is to relabel $(f_i)_{I\in I\setminus I_0}$ as $(e_j)_{j=1,\dots,n}$.

6.3 This section contains the first part of the proof of the Rohlin theorem. We show that almost all the space can be almost filled up with mutually orthogonal projections, each of them suitable to become a tower basis for a Rohlin tower.

Let us come back to the notation used in the statement of Theorem 6.1. We shall work in the relative commutant in M of a countably generated α-invariant sub W^*-algebra N of M_ω. For each $g,h \in G$ there exists a unitary $u_{g,h} \in M_\omega$ such that $\alpha_g\alpha_h = \mathrm{Ad}\, u_{g,h}\alpha_{gh}$. We may assume that N contains all $u_{g,h}$ and thus that $\alpha|N'\cap M_\omega$ is an action. Since α is strongly free, $\alpha|N'\cap M_\omega$ is free; moreover $N'\cap M_\omega$ is finite and the trace ϕ_ω (which depends only on $\phi|Z(M)$) is α-invariant.

LEMMA. *Let* $\delta > 0$ *and let* K *be a finite nonempty subset of* G, $1 \notin K$. *Then there exists a partition of unity* $(e_i)_{i=0,\dots,q}$ *in* $N'\cap M_\omega$ *such that*

(1) $|e_0|_\phi \leqslant \delta$

(2) $e_i\alpha_g(e_i) = 0$ *for* $1 \leqslant i \leqslant q$, $g\in K$.

<u>Proof</u>. <u>Step A</u>. Let $\gamma > 0$ and $f \in \mathrm{Proj}(N'\cap M_\omega)$, $f \neq 0$. We show that there exists $f' \in \mathrm{Proj}(N'\cap M_\omega)$, $0 \neq f' \leqslant f$, such that

$$|f'\alpha_g(f')|_\phi \leqslant 2\gamma|f'|_\phi \quad , \qquad g\in K .$$

Let $\bar N$ be the smallest α-invariant subalgebra of M_ω, containing both N and f. Then α is free on $\bar N'\cap M_\omega$, and by Corollary 6.2 we may choose a partition of unity $(f_i)_{i=0,\dots,m}$ in $\bar N'\cap M_\omega$ such that

(3) $|f_0|_\phi \leqslant \frac{1}{2}|f|_\phi$

(4) $\sum_{g\in K} |f_i\alpha_g(f_i)|_\phi < \gamma|f|_\phi\, |f_i|_\phi$, $i = 1,\dots,m$.

Let $\bar f_i = f_i f \in \mathrm{Proj}(N'\cap M_\omega)$ and suppose that for each $i = 1,\dots,m$

$$\sum_{g\in K} |\bar{f}_i\alpha_g(\bar{f}_i)|_\phi \geq 2\gamma|\bar{f}_i|_\phi \quad .$$

Then the assumed commutativity relations together with (3) yield

$$\sum_{i=1}^m \sum_{g\in K} |f_i\alpha_g(f_i)|_\phi = \phi_\omega\Big(\sum_{i=1}^m \sum_{g\in K} |f_i\alpha_g(f_i)|\Big)$$

$$\geq \phi_\omega\Big(\sum_{i=1}^m \sum_{g\in K} |f_i\alpha_g(f_i)|\,|f\alpha_g(f)|\Big)$$

$$= \sum_{i=1}^m \sum_{g\in K} |\bar{f}_i\alpha_g(\bar{f}_i)|_\phi \geq 2\gamma \sum_{i=1}^m |\bar{f}_i|_\phi$$

$$= 2\gamma\phi_\omega((1-f_0)f) \geq 2\gamma(|f|_\phi - |f_0|_\phi) \geq \gamma|f|_\phi \, .$$

On the other hand, from (4),

$$\sum_{i=1}^m \sum_{g\in K} |f_i\alpha_g(f_i)|_\phi < \gamma|f|_\phi \sum_{i=1}^m |f_i|_\phi \leq \gamma|f|_\phi \quad .$$

The contradiction thus obtained shows that for some $i \in \{1,\dots,m\}$,

$$\sum_{g\in K} |\bar{f}_i\alpha_g(\bar{f}_i)|_\phi < 2\gamma|\bar{f}_i|_\phi$$

and thus we may take $f' = \bar{f}_i$.

<u>Step B</u>. We show that for any $f \in \mathrm{Proj}(N' \cap M_\omega)$ and any $\gamma > 0$ there exists $e \in \mathrm{Proj}(N' \cap M_\omega)$ with

(5) $e \leq f$

(6) $|e\alpha_g(e)|_\phi \leq \gamma|e|_\phi$, $g \in K$

(7) $|e|_\phi \leq (1+|K|)^{-1}|f|_\phi$.

The family of projections $e \in N' \cap M_\omega$ satisfying (5) and (6) is nonvoid and well ordered, so let e be maximal with these properties. We show that e also satisfies

(8) $e \vee \big(\bigvee_{g\in K} \alpha_g(e)\big) \vee (1-f) = 1$.

If not, let e' be a nonzero projection in $N' \cap M_\omega$ orthogonal to the left member of (8). By Step A there exists a nonzero projection e'' in $N' \cap M_\omega$, $e'' \leq e'$, with $|e''\alpha_g(e'')|_\phi \leq \delta_1|e''|_\phi$, $g \in K$. We have $e'' \leq f$ and $e''\alpha_g(e) = 0$ for $g \in K$, hence $e+e''$ satisfies (5) and (6). The assumed maximality of e is contradicted, and thus (8) is proved.

From (8) we get

$$1 = \Big|e \vee \Big(\bigvee_{g\in K} \alpha_g(e)\Big) \vee (1-f)\Big|_\phi \leq |e|_\phi + \sum_{g\in K} |\alpha_g(e)|_\phi + |1-f|_\phi$$

$$= 1 - |f|_\phi + (1+|K|)|e|_\phi$$

and (7) is proved.

Step C. Let $q \in \mathbb{N}$ be such that $(1-(1+|K|)^{-1})q < \delta$. We now prove a weaker version of the lemma, showing that for any $\gamma > 0$ there exists a partition of unity $(e_i)_{i=0,\dots,q}$ in $N' \cap M_\omega$ such that

$$|e_0|_\phi < \delta$$

$$(9) \quad |e_i \alpha_g(e_i)|_\phi < \gamma |e_i|_\phi, \quad i=1,\dots,q; \quad g \in K.$$

Let us take $f_1 = 1$ and construct successively for $k=1,\dots,q$, according to Step B, projections e_k and f_{k+1} in $N' \cap M_\omega$ such that $e_k \leqslant f_k$, $f_{k+1} = f_k - e_k$, and

$$|e_k \alpha_g(e_k)|_\phi \leqslant \gamma |e_k|_\phi \qquad g \in K$$

$$|e_k|_\phi \geqslant (1+|K|)^{-1} |f_k|_\phi .$$

We have $|f_{k+1}|_\phi \leqslant (1-(1+|K|)^{-1})|f_k|_\phi$ for all k, thus $|f_{q+1}|_\phi \leqslant (1-(1+|K|)^{-1})^q \leqslant \delta$ and letting $e_0 = f_{q+1}$, Step C is proved.

Step D. Since γ can be taken arbitrarily small, and q does not depend on it, we may apply the Index Selection Trick **5.5** to the projections $e_0,\dots,e_q$ obtained above to make $\gamma = 0$ in (9) and thus prove the lemma. Let us describe this procedure in detail.

For any natural $n \geqslant 1$, let us choose a family $(e_k^{(n)})$, $k=0,\dots,q$ of projections in $N' \cap M_\omega$ with

$$\sum_k e_k^{(n)} = 1$$

$$|e_k^{(0)}|_\phi \leqslant \delta$$

$$|e_k^{(n)} \alpha_g(e_k^{(n)})|_\phi \leqslant \frac{1}{n}, \qquad k=1,\dots,q; \quad g \in K .$$

Let $(U_m)_{m \in N}$ be unitaries generating N, and let $\mathcal{A} = \{\alpha_g | g \in G\} \cup \{\mathrm{Ad}\, u_m | m \in \mathbb{N}\} \subset \mathrm{Aut}\, M_\omega$. Let $\tilde{e}_k = (e_k^{(n)})_n \in \ell^\infty(\mathbb{N}, M_\omega)$, $k=0,\dots,q$, and let $\mathcal{C}$ be a separable sub C^*-algebra of $\ell^\infty(\mathbb{N}, M_\omega)$ which contains all the projections $\tilde{e}_k$ and which is kept globally invariant by the automorphisms in $\mathcal{A}$ (acting term by term on $\ell^\infty(\mathbb{N}, M_\omega)$).

Let $\Psi: \mathcal{C} \to M_\omega$ be the homomorphism yielded by the Index Selection Trick. If $e_k = \Psi(\tilde{e}_k) \in M_\omega$, $k=0,\dots,q$ then e_k are projections of sum 1, and satisfy

$$|e_0|_\phi = \phi_\omega(e_0) = \lim_{n\to\infty} \phi_\omega(e_0^{(n)}) \leqslant \delta$$

and similarly

$$|e_k \alpha_g(e_k)|_\phi = \lim_{n\to\infty} |e_k^{(n)} \alpha_g(e_k^{(n)})|_\phi = 0, \quad k=1,\dots,q, \ g \in K.$$

We also have for all $m \in \mathbb{N}$

$$\operatorname{Ad} u_m(e_k) = \operatorname{Ad} u_m(\Psi(\tilde{e}_k)) = \Psi((\operatorname{Ad} u_m(e_k^{(n)}))_n)$$
$$= \Psi(\tilde{e}_k) = e_k \,, \qquad k = 0,\dots,q$$

and thus $e_k \in N' \cap M_\omega$. The lemma is proved.

In the following we shall apply the Index Selection Trick several times in the same manner as above, in order to get genuine equalities in M^ω or M_ω out of approximate ones.

6.4 We begin the second part of the proof of the Rohlin theorem by associating to a family $E = (E_{i,k})$ of mutually orthogonal projections in M_ω, indexed by $i \in I = \{1,\dots,N\}$ and $k \in K_i$ ($K_1,\dots,K_N$ being the ε-paving subsets of G in the statement of 6.1) the following numbers

$$a_E = \sum_i |K_i|^{-1} \sum_{k,\ell \in K_i} |\alpha_{k\ell^{-1}}(E_{i,\ell}) - E_{i,k}|_\phi$$

$$b_E = \sum_{i,k} |E_{i,k}|_\phi$$

and for $g \in G$

$$c_{g,E} = \sum_{i,j} \sum_{k,\ell} |[\alpha_g(E_{i,k}), E_{j,\ell}]|_\phi \ .$$

Recall that N is a countably generated sub W^*-algebra of M_ω, α-invariant and such that $\alpha|N' \cap M_\omega$ is an action.

LEMMA. *Let* $E = (E_{i,k})$ *be a family of mutually orthogonal projections in* $N' \cap M_\omega$. *Let* $\delta > 0$ *and* $A \subset\subset G$ *be given and suppose that* $0 < \varepsilon < 1/16$.

If $b_E < 1 - \varepsilon^{1/2}$ *then there is a family* $E' = (E'_{i,k})$ *of mutually orthogonal projections in* $N' \cap M_\omega$ *such that*

(1) $0 < \varepsilon \sum_{i,k} |E_{i,k} - E'_{i,k}|_\phi \leqslant b_{E'} - b_E$

(2) $a_{E'} - a_E \leqslant 3\varepsilon^{1/2}(b_{E'} - b_E)$

(3) $c_{g,E'} - c_{g,E} \leqslant 3\delta\varepsilon^{-1}(b_{E'} - b_E)$ *for* $g \in A$.

Proof. The idea of the proof of (1) and (2) is the following. If all $E_{i,k}$ are zero, then a tower $(E'_{i,k})$ is supplied by the previous lemma. If not, we choose among the projections yielded by that lemma a tower base and then construct a tower $(f_{i,k})$, such that all $f_{i,k}$

commute with all $E_{i,k}$; then with $\tilde{f} = \sum_{i,k} f_{i,k}$ we take $E'_{i,k} = E_{i,k}(1-\tilde{f}) + f_{i,k}$. In (1) it is required that E' be significantly larger than E, i.e. that $E_{i,k}\tilde{f}$ be small with respect to $\tilde{f}$; this is achieved by an adequate choice of the tower basis. In view of (2) we should care that a_E, which measures the failure of $(E_{i,k})$ to be equivariant, does not increase too much. The only problem is the fact that we alter the old tower.

If $\tilde{f}$ was α-invariant then cutting with $1-\tilde{f}$ would not affect a_E. We approximate this by taking a tower indexed by a very large subset K' of G; such a tower has a very good global invariance, and subsequently we regroup its projections to get the tower $(f_{i,k})$ indexed by $K_1,\dots,K_N$.

For $G = \mathbb{Z}$, $I = \{1\}$ and $K_1 = \{1,2,\dots,p\} \subset \mathbb{Z}$, a typical picture would be the following:

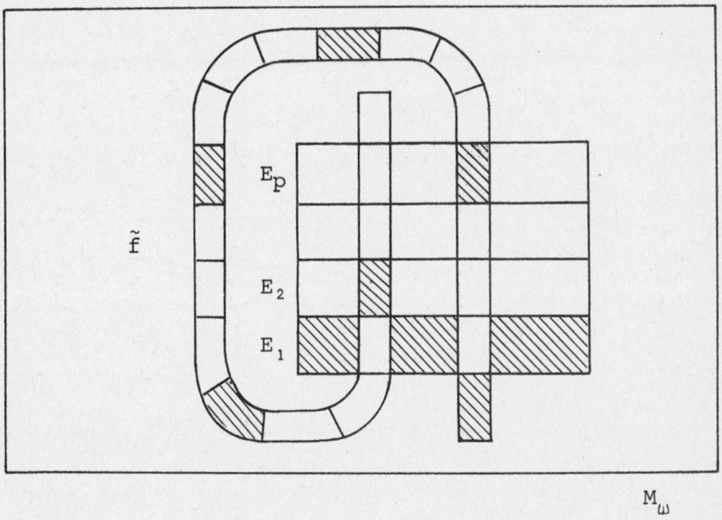

In the figure, E'_1, the new tower basis, is shaded and is obtained from the old E_1 by taking out $E_1\tilde{f}$ and adding the basis of $\tilde{f}$ rearranged as a K_1 tower, i.e. the dark parts of $\tilde{f}$. The projection $\tilde{f}$ has a very large invariance degree with respect to α.

Let us begin the proof. Since we have assumed $0 < \varepsilon < 1/16$, there exists ε_1, $0 < \varepsilon_1 < \varepsilon$, such that

(4) $b_E < (1-\varepsilon^{1/2})(1-\varepsilon_1)$

(5) $2(\varepsilon_1 + \varepsilon(1-\varepsilon)^{-1})(\varepsilon^{1/2} - \varepsilon)^{-1} < 3\varepsilon^{1/2}$.

We may suppose that $\delta < (\sum_i |K_iK_i^{-1}|)^{-1}\varepsilon_1$ and that $A \supset \bigcup_i K_iK_i^{-1}$.

<u>Step A</u>. Let K' be a (δ, A)-invariant subset of G, which is ε-paved by $K_1,\dots,K_N$. According to Lemma 6.3 (with $(K')^{-1}K'\setminus\{1\}$

standing for K), choose a partition of unity $(e_i)_{i=0,\dots,q}$ in $N' \cap M_\omega$ with

$$|e_0|_\phi < \varepsilon_1$$

$$\alpha_g(e_j)\alpha_h(e_j) = 0\,, \quad j=1,\dots,q, \quad g,h \in K'\,, \quad g \neq h$$

$$[e_j, \alpha_g(E_{i,k})] = 0 \quad \text{for all } j,i,k,g\,.$$

Letting

$$x = |K'|^{-1} \sum_{g\in K'} \alpha_g^{-1}\Big(\sum_{i,k} E_{i,k}\Big)$$

we have

$$|x|_\phi = \phi_\omega(x) = \phi_\omega\Big(\sum_{i,k} E_{i,k}\Big) = b_E \quad ,$$

and moreover x commutes with all e_i.

There exists $j \in \{1,\dots,q\}$ such that

$$|e_j x|_\phi \leq (1-\varepsilon^{\frac{1}{2}})|e_j|_\phi\,.$$

If not, adding the opposite inequalities for $j=1,\dots,q$ we infer

$$b_E = |x|_\phi \geq |(1-e_0)x|_\phi > (1-\varepsilon^{\frac{1}{2}})(1-|e_0|_\phi) > (1-\varepsilon^{\frac{1}{2}})(1-\varepsilon_1)$$

and thus contradict the hypothesis (4).

We let $f = e_j$, $f' = \sum_{g\in K'} \alpha_g(f)$ and $\rho = |f'|_\phi$. Then $|fx|_\phi \leq (1-\varepsilon^{\frac{1}{2}})|f|_\phi$ and so

$$\begin{aligned}(6) \qquad |f' \sum_{i,k} E_{i,k}|_\phi &= \sum_{g\in K'} |\alpha_g(f) \sum_{i,k} E_{i,k}|_\phi \\ &= \sum_{g\in K'} |f\alpha_g^{-1}\Big(\sum_{i,k} E_{i,k}\Big)|_\phi = |K'|\,|fx|_\phi \\ &\leq (1-\varepsilon^{\frac{1}{2}})|K'|\,|f|_\phi = (1-\varepsilon^{\frac{1}{2}})\rho\,.\end{aligned}$$

We assumed that (K_i), $i \in I$, ε-pave K'. Hence there are subsets (L_i), $i \in I$ of G and $K'_{i,\ell} \subseteq K_i$, $i \in I$, $\ell \in L_i$, such that if $\tilde{K} = \bigcup_i K_i L_i$, then

$$(7) \qquad \begin{aligned} &|K'_{i,\ell}| \geq (1-\varepsilon)|K_i| \\ &|K' \setminus \tilde{K}| \leq \varepsilon |K'|\,. \end{aligned}$$

Let us now define for $i \in I$, $k \in K_i$

$$S_{i,k} = \{k\ell \mid \ell \in L_i,\ K'_{i,\ell} \ni k\}\,, \qquad S_i = \bigcup_k S_{i,k}\,.$$

Accordingly, let us take for $i \in I$, $k \in K_i$

$$f_{i,k} = \sum_{g \in S_{i,k}} \alpha_g(f)$$

$$f_i = \sum_k f_{i,k}$$

$$\tilde{f} = \sum_k f_i \quad .$$

Then $\tilde{f} = \sum_{g \in \tilde{K}} \alpha_g(f) \leq f' = \sum_{g \in K'} \alpha_g(f)$, and by (7) we have

$$|\tilde{f}|_\phi = |\tilde{K}|\,|f|_\phi \geq (1-\varepsilon)|K'|\,|f|_\phi = (1-\varepsilon)|f'|_\phi \,,$$

that is,

$$(8) \qquad |\tilde{f}|_\phi \geq (1-\varepsilon)\rho \quad .$$

Let $K_\Delta = \bigcup_{i \in I} \bigcup_{k,\ell \in K_i} (K' \Delta k\ell^{-1} K')$. Since for each i, K' is $(\varepsilon_1 |K_i K_i^{-1}|^{-1}, K_i K_i^{-1})$-invariant, we infer $|K_\Delta| \leq 2\varepsilon_1 |K'|$ and thus if we let $f_\Delta = \sum_{g \in K_\Delta} \alpha_g(f)$ then

$$(9) \qquad |f_\Delta|_\phi \leq 2\varepsilon_1 \rho \quad .$$

We are now in a position to define the family $E' = (E'_{i,k})$ by taking

$$E'_{i,k} = (1-f')E_{i,k} + f_{i,k} \,, \qquad i \in I \,, \quad k \in K_i \quad .$$

The amount of modifications from E to E' is estimated by

$$(10) \qquad \sum_{i,k} |E'_{i,k} - E_{i,k}|_\phi \leq |f'|_\phi + \sum_{i,k} |f_{i,k}|_\phi$$
$$= |f'|_\phi + |\tilde{f}|_\phi \leq 2\rho \quad .$$

In view of (8) and (6) this gives

$$(11) \qquad b_{E'} = |\sum_{i,k} E'_{i,k}|_\phi = |\sum_{i,k} E_{i,k}|_\phi + |\tilde{f}|_\phi - |\sum_{i,k} E_{i,k} f'|_\phi$$
$$\geq b_E + (1-\varepsilon)\rho - (1-\varepsilon^{\frac{1}{2}})\rho \geq b_E + (\varepsilon^{\frac{1}{2}} - \varepsilon)\rho \geq b_E + 2\varepsilon\rho$$

and thus (10) yields

$$b_{E'} - b_E \geq \varepsilon \sum_{i,k} |E'_{i,k} - E_{i,k}|_\phi \quad .$$

We have proved the statement (1) in the conclusion of the lemma.

Step B. Let us now prove the second part of the lemma, concerning the equivariance of the Rohlin towers. If $i \in I$ and $k,m \in K_i$ we infer

$$(12) \quad |\alpha_{km^{-1}}(E'_{i,m}) - E'_{i,k}|_\phi$$

$$\leq |(\alpha_{km^{-1}}(E'_{i,m}) - E_{i,k})(1-\alpha_{km^{-1}}(f'))|_\phi$$

$$+ |E_{i,k}(f' - \alpha_{km^{-1}}(f'))|_\phi + |\alpha_{km^{-1}}(f_{i,m}) - f_{i,k}|_\phi$$

$$\leq |\alpha_{km^{-1}}(E_{i,m}) - E_{i,k}|_\phi + |E_{i,k}f_\Delta|_\phi + |(km^{-1}S_{i,m})\Delta S_{i,k}|\,|f|_\phi .$$

For each $i \in I$ we have

$$\sum_{k,m\in K_i} |(km^{-1}S_{i,m})\Delta S_{i,k}| \leq \sum_{k,m\in K_i} (|(m^{-1}S_{i,m})\Delta L_i| + |(k^{-1}S_{i,m})\Delta L_i|)$$

$$= 2|K_i| \sum_{k\in K_i} |(k^{-1}S_{i,k})\Delta L_i|$$

$$= 2|K_i| \sum_{k\in K_i} |\{\ell \in L_i | K'_{i,\ell} \not\ni k\}|$$

$$= 2|K_i| \sum_{\ell\in L_i} |\{k \in K_i | k \notin K'_{i,\ell}\}|$$

$$\leq 2\varepsilon|L_i|\,|K_i|^2 \leq 2\varepsilon(1-\varepsilon)^{-1}|K_i| \sum_{\ell\in L_i} |K'_{i,\ell}|$$

$$= 2\varepsilon(1-\varepsilon)^{-1}|K_i|\,|K_iL_i| .$$

If we take this into (11) and sum up, we obtain

$$a_{E'} = \sum_i |K_i|^{-1} \sum_{k,m} |\alpha_{km^{-1}}(E'_{i,m}) - E'_{i,k}|_\phi$$

$$\leq \sum_i |K_i|^{-1} \sum_{k,m} |\alpha_{km^{-1}}(E_{i,m}) - E_{i,k}|_\phi + |f_\Delta|_1 + 2\varepsilon(1-\varepsilon)^{-1}|K'|\,|f|_\phi$$

$$= a_E + |f_\Delta|_1 + 2\varepsilon(1-\varepsilon)^{-1}\rho .$$

In view of (9), (11), and our assumption (5) on ε, this yields

$$a_{E'} \leq a_E + 2\varepsilon_1\rho + 2\varepsilon(1-\varepsilon)^{-1}\rho$$

$$\leq a_E + (2\varepsilon_1 + 2\varepsilon(1-\varepsilon)^{-1})(\varepsilon^{1/2} - \varepsilon)^{-1}(b_{E'} - b_E)$$

$$\leq a_E + 3\varepsilon^{1/2}(b_{E'} - b_E)$$

and the proof is finished.

Step C. We now prove the third statement of the lemma, concerning the mutual approximate commutation of projections of the form $\alpha_g(E'_{i,k})$. Since the tower $(f_{i,k})$ commutes with all $\alpha_g(E_{j,\ell})$, the only problem that remains is $(f_{i,k})$ itself. The projections $f_{i,k}$ are sums of mutually orthogonal projections of the tower $(\alpha_m(f))_{m\in K'}$. Since K' is almost invariant to $g\in A$, $\alpha_g(f_{i,k})$ will be approximately equal to a part of this tower too; but the projections $(\alpha_m(f))_{m\in K'}$ mutually commute.

For $g\in A$, $i\in I$ and $k\in K_i$ we have

$$\alpha_g(f_{i,k})f' = \sum_h \alpha_h(f)$$

where $h\in(gS_{i,k})\cap K'$. Hence

$$\sum_{i,k} |\alpha_g(f_{i,k})(1-f')|_\phi \leqslant |g(\bigcup_{i,k} S_{i,k})\setminus K'|\,|f|_\phi$$

$$\leqslant |gK'\setminus K'|\,|f|_\phi \leqslant \delta|K'|\,|f|_\phi = \delta\rho$$

since K' was assumed (δ,A) invariant. We also infer

$$\sum_{i,k} |\alpha_g(E_{i,k}(1-f')) - \alpha_g(E_{i,k})(1-f')|_\phi$$

$$\leqslant |\alpha_g(f')-f'|_\phi \leqslant |gK'\Delta K'|\,|f|_\phi \leqslant 2\delta\rho \quad .$$

Since $E'_{i,k} = E_{i,k}(1-f')+f_{i,k}$ we obtain

$$\sum_{i,k}\sum_{j,\ell} |[\alpha_g(E'_{i,k}),E'_{j,\ell}]|_\phi$$

$$-\sum_{i,k}\sum_{j,\ell} |[\alpha_g(E_{i,k})(1-f')+\alpha_g(f_{i,k})f',\ E_{j,\ell}(1-f')+f_{j,\ell}]|_\phi$$

$$\leqslant 2(\delta\rho+2\delta\rho) = 6\delta\rho \quad .$$

Since $\alpha_g(f_{i,k})f'_\ell$ and $f_{j,\ell}$ are sums of mutually orthogonal projections from the tower $(\alpha_h(f))_{h\in K'}$, they commute with each other and with the tower E. We thus have

$$c_{g,E'} \leqslant 6\delta\rho + \sum_{v,k}\sum_{j,\ell} |[\alpha_g(E_{i,k}),E_{j,\ell}](1-f')|_\phi$$

$$\leqslant 6\delta\rho + c_{g,E} \leqslant c_{g,E} + 3\delta\varepsilon^{-1}(b_{E'}-b_E)$$

and the proof of (3) is also finished. The lemma is proved.

6.5 The Rohlin Theorem is obtained now from the preceding lemma by a maximality argument. Let us keep $A \subset\subset G$ and $\delta > 0$ fixed. Let $\mathcal{E}$ be the set of families $E = (E_{i,k})_{i \in I, k \in K_i}$ of mutually orthogonal projections in $N' \cap M_\omega$ (see **6.3** and **6.4**) satisfying

(1) $a_E \leqslant 3\varepsilon^{\frac{1}{2}} b_E$

(2) $c_{g,E} \leqslant 3\delta\varepsilon^{-1} b_E$, $g \in A$.

$\mathcal{E}$ is nonvoid since it contains the null family. We order $\mathcal{E}$ by letting $E \leqslant E'$ if either $E = E'$ or the conclusion of Lemma 6.4 holds for E and E'. For any totally ordered subset of $\mathcal{E}$, the map $E \to b_E$ is, by 6.4(1), an order ismorphism with a subset of the interval $[0,1] \subset \mathbb{R}$, and again by 6.4(1) for any increasing net in a totally ordered subset of $\mathcal{E}$, the projections $E_{i,k}$ will converge in the s*-topology to the components of an element of $\mathcal{E}$; hence $\mathcal{E}$ is inductively ordered and, by the Zorn lemma, has a maximal element E^0. Lemma 6.4 shows that E^0 satisfies $b_{E^0} \geqslant 1 - \varepsilon^{\frac{1}{2}}$, where (1) and (2) above come from 6.4(2) and 6.4(3) respectively and so, letting $E^0_0 = 1 - \sum_{i,k} E^0_{i,k}$, we have $|E^0_0|_\phi \leqslant \varepsilon^{\frac{1}{2}}$.

To get rid of E^0_0, we choose some arbitrary $\bar{i} \in I$ and $\bar{k} \in K_{\bar{i}}$ and define $E_{i,k} = E^0_{i,k}$ if $(i,k) \neq (\bar{i},\bar{k})$ and $E_{\bar{i},\bar{k}} = E^0_{\bar{i},\bar{k}} + E^0_0$. This way, $E_{i,k}$ is a partition of unity and it satisfies

(3) $a_E \leqslant 5\varepsilon^{\frac{1}{2}}$

(4) $c_{g,E} \leqslant 3c_{g,E^0} \leqslant 9\delta\varepsilon^{-1}$, $g \in A$

since for $g \in A$, the new terms in $c_{g,E}$ are estimated by

$$\sum_{j,\ell} |[\alpha_g(E^0_0), E_{j,\ell}]|_\phi = \sum_{j,\ell} |[1 - \sum_{i,k} \alpha_g(E^0_{i,k}), E^0_{j,\ell}]|_\phi \leqslant c_{g,E}$$

and similarly, $\sum_{j,\ell} |[\alpha_g(E^0_{j,\ell}), E^0_0]|_\phi \leqslant c_{g,E}$.

For any given $\delta > 0$ and $A \subset\subset G$ we may thus find a partition of unity $E = (E_{i,k})_{i,k}$ in $N' \cap M_\omega$ satisfying (3) and (4). We may apply the Index Selection Trick the same way as we did in **6.3**, Step D, for $\delta \searrow 0$ and $A \nearrow G$, in order to obtain (4) with $\delta = 0$ and $A = G$. Thus Theorem 6.1 is proved.

6.6 Suppose that a discrete amenable group G acts on M like in the statement of Theorem 6.1, and let H be a normal subgroup of G. Then the Rohlin Theorem holds for the action that the quotient G/H induces on the fixed point algebra $(M_\omega)^H$. To avoid technical complication we prove the result only in the case when the subgroup is a direct summand, which is what we need in the sequel, but the proof extends along the same lines to the general case. For simplicity we also assume that the algebra M is a factor, and we denote by τ the canonical trace τ_ω on M_ω.

THEOREM (Relative Rohlin Theorem). *Let* G *and* $\bar{G}$ *be discrete countable amenable groups, and let* M *be a factor with separable predual. Let* $\theta: G \times \bar{G} \to \operatorname{Aut} M_\omega$ *be a crossed action, which is semiliftable and strongly free. Let* $\varepsilon > 0$ *and let* $(K_i)_{i \in I}$ *be an* ε*-paving family of subsets of* G.

Then there exists a partition of unity $(E_{i,k})$, $i \in I$; $k \in K_i$ *in* M_ω *such that if* $\alpha_g = \theta_{(g,1)}$ *and* $\beta_g = \theta_{(1,g)}$ *then*

(1) $\sum_k |K_i|^{-1} \sum_{k,\ell} |\alpha_{k\ell^{-1}}(E_{i,\ell}) - E_{i,k}|_\tau \leqslant 16\varepsilon$

(2) $\beta_g(E_{i,k}) = E_{i,k}, \quad g \in G, \quad i \in I, \quad k \in K_i$

(3) $[E_{i,k}, \alpha_g(E_{j,\ell})] = 0$ *for all* g,i,k,j,ℓ

(4) $\theta_{(g,h)}\theta_{(\ell,m)}(E_{i,k}) = \theta_{(g\ell,hm)}(E_{i,k})$ *for all* g,h,ℓ,m,i,k.

Moreover $(E_{i,k})$ *can be chosen in the relative commutant in* M_ω *of any given countable subset of* M_ω.

Remark. The estimate (1) above improves (1) of **6.1** (if we take $\bar{G}$ trivial), being linear in ε.

Proof. The idea of the proof is to take Rohlin towers indexed by products of (very large) sets in $G \times \bar{G}$, and then sum after the $\bar{G}$ coordinate.

Step A. We assume $0 < \varepsilon < 1/16$ and choose $\bar{A} \subset\subset \bar{G}$. We prove first that the theorem holds with (1) and (2) replaced by

(1') $\sum_i |K_i|^{-1} \sum_{k,\ell} |\alpha_{k\ell^{-1}}(E_{i,\ell}) - E_{i,k}|_\tau \quad 16\varepsilon^{1/2}$

(2') $\sum_{i,k} |\beta_g(E_{i,k}) - E_{i,k}|_\tau \quad 34\varepsilon^{1/2}, \quad g \in \bar{A}$.

Let $(\bar{K}_i)_{i \in \bar{I}}$ be an ε-paving family of subsets of $\bar{G}$, all of them $(\varepsilon, \bar{A})$-invariant. It is easy to see that the family $(K_i \times \bar{K}_j)_{i,j}$ of subsets of $G \times \bar{G}$ 2ε-paves any subset of $G \times \bar{G}$ of the form $S \times \bar{S}$ if $S \subset\subset G$ and $\bar{S} \subset\subset \bar{G}$ are invariant enough. This doesn't imply that $(K_i \times \bar{K}_j)_{i,j}$ is a 2ε-paving family for $G \times \bar{G}$, but in the proof of the Rohlin Theorem we need only the fact that for any invariance degree, the given family of subsets of the group ε-paved some subset (and not necessarily all subsets) of the group having that invariance degree. We may thus apply the Rohlin Theorem to obtain a partition of unity $(F_{(i,\bar{i}),(k,\bar{k})})$ in M_ω, with $(i,\bar{i}) \in I \times \bar{I}$ and $(k,\bar{k}) \in K_i \times \bar{K}_{\bar{i}}$, such that

$$(5)\quad \sum_{i,\bar{i}} |K_i|^{-1}|\bar{K}_{\bar{i}}|^{-1} \sum_{k,\bar{k},\ell,\bar{\ell}} |\alpha_{k\ell^{-1}}\beta_{\bar{k}\bar{\ell}^{-1}}(F_{(i,\bar{i}),(\ell,\bar{\ell})}) - F_{(i,\bar{i}),(k,\bar{k})}|_\tau$$

$$\leqslant 5 \times (2\varepsilon)^{\frac{1}{2}} \leqslant 8\varepsilon^{\frac{1}{2}}$$

$$(6)\quad [\alpha_g\beta_{\bar{g}}(F_{(i,\bar{i}),(k,\bar{k})}), F_{(j,\bar{j}),(\ell,\bar{\ell})}] = 0 \quad \text{for all } \bar{g}, g, i, \bar{i}, k, \bar{k}, j, \bar{j}, \ell, \bar{\ell}$$

$$(7)\quad \alpha_g\beta_{\bar{g}}\alpha_h\beta_{\bar{h}}(F_{(i,\bar{i}),(k,\bar{k})}) = \alpha_{gh}\beta_{\bar{g}\bar{h}}(F_{(i,\bar{i}),(k,\bar{k})}) \quad \text{for all } g,\bar{g},h,\bar{h},i,\bar{i},k,\bar{k}.$$

Let us take $E_{i,k} = \sum_{\bar{i},\bar{k}} F_{(i,\bar{i}),(k,\bar{k})}$, $i \in I$, $k \in K_i$, $\bar{i} \in \bar{I}$, $\bar{k} \in \bar{K}_{\bar{i}}$.
For any $i \in I$ and $k, \ell \in K_i$ we infer

$$\alpha_{k\ell^{-1}}(E_{i,\ell}) - E_{i,k} = \sum_{\bar{i},\bar{\ell}}\sum \left(\alpha_{k\ell^{-1}}(F_{(i,i),(\ell,\ell)}) - F_{(i,\bar{i}),(\ell,\bar{\ell})}\right)$$

$$= -\alpha_{k\ell^{-1}}\left(\sum_{\bar{i}} |\bar{K}_{\bar{i}}|^{-1} \sum_{\bar{\ell},\bar{m}} \left(\alpha_{\ell m^{-1}}\beta_{\bar{\ell}\bar{m}^{-1}}(F_{(i,\bar{i}),(m,\bar{m})}) - F_{(i,\bar{i}),(\ell,\bar{\ell})}\right)\right)$$

$$+ \sum_{\bar{i}} |\bar{K}_{\bar{i}}|^{-1} \sum_{\bar{\ell},\bar{m}} \left(\alpha_{\ell m^{-1}}\beta_{\bar{\ell}\bar{m}^{-1}}(F_{(i,\bar{i}),(m,\bar{m})}) - F_{(i,\bar{i}),(k,\bar{\ell})}\right).$$

Hence from (5)

$$\sum_i |K_i|^{-1} \sum_{k,\ell} |\alpha_{k\ell^{-1}}(E_{i,\ell}) - E_{i,k}|_\tau \leqslant 2 \times 8\varepsilon^{\frac{1}{2}} = 16\,^{\frac{1}{2}}$$

and (1') is proved.

For $g \in A$ we have

$$\sum_{i,k} (\beta_g(E_{i,k}) - E_{i,k}) = \Sigma_1 + \beta_g(\Sigma_2) + \Sigma_3$$

where

$$\Sigma_1 = \sum_{i,i}\sum_{k,k} (\beta_g(F_{(i,i),(k,k)}) - F_{(i,\bar{i}),(k,g\bar{k})})$$

$$\Sigma_2 = \sum_{i,\bar{i}}\sum_{k,\bar{\ell}} F_{(i,\bar{i}),(k,\bar{\ell})}$$

$$\Sigma_3 = \sum_{i,\bar{i}}\sum_{k,\bar{m}} F_{(i,\bar{i}),(k,\bar{m})}$$

and the sums were done for $i \in I$, $\bar{i} \in \bar{I}$, $k \in K_i$, $\bar{k} \in \bar{K}_{\bar{i}} \cap g^{-1}\bar{K}_{\bar{i}}$, $\bar{\ell} \in \Delta_{\bar{i}} = \bar{K}_{\bar{i}} \setminus g^{-1}\bar{K}_{\bar{i}}$ and $m \in g\Delta_{\bar{i}}$.

From the assumed $(\varepsilon,\bar{A})$-invariance of $\bar{K}_{\bar{i}}$, we infer $|\Delta_{\bar{i}}| \leqslant \varepsilon|\bar{K}_{\bar{i}}|$ for all $\bar{i}$, hence with the global estimates in Corollary 6.1 corresponding to (4) above, we infer

$$|\Sigma_1|_\tau \leq 2\times 8\varepsilon^{\frac{1}{2}} = 16\varepsilon^{\frac{1}{2}}$$

$$|\Sigma_2|_\tau \leq \varepsilon + 8\varepsilon^{\frac{1}{2}} \leq 9\varepsilon^{\frac{1}{2}}$$

$$|\Sigma_3|_\tau \leq \varepsilon + 8\varepsilon^{\frac{1}{2}} \leq 9\varepsilon^{\frac{1}{2}}$$

and thus for any $g \in \bar{A}$,

$$\sum_{i,k} |\beta_g(e_{i,k}) - E_{i,k}|_\tau \leq |\Sigma_1|_\tau + |\Sigma_2|_\tau + |\Sigma_3|_\tau \leq 34\varepsilon^{\frac{1}{2}}$$

and (2') is proved too.

Step B. We want to obtain the estimate (2') with an arbitrarily small constant. We do this by starting with better paving subsets $(K_j')_j$ of G and then come back, by means of the Paving Theorem, from (K_j') towers to (K_i) towers.

Recall that we are given $\varepsilon > 0$ and the ε-paving family $(K_i)_{i\in I}$ of subsets of G. Let $\delta > 0$ and $\bar{A} \subset\subset \bar{G}$. Let us use Corollary 3.3 the same way as in the construction of the Paving Structure 3.4, to obtain a system $(K_j')_{j\in J}$ of finite subsets of G, δ-paving G, and finite subsets $(L_{i,j})_{i\in I, j\in J}$ of G with

$$|K_j'| = \sum_i |K_i||L_{i,j}| \quad , \quad j \in J ,$$

and such that the subsets

$$(8) \quad \tilde{K}_{i,j} = \{h \in K_j' \mid \text{there are unique } (\tilde{i},k,\ell) \in \sum_{\tilde{i}\in I} K_{\tilde{i}} \times L_{\tilde{i},j} \text{ with } h = k\ell \text{ and for these } \tilde{i}=i\}$$

satisfy

$$|\tilde{K}_{i,j}| \geq (1-4\varepsilon)|K_i||L_{i,j}| \quad .$$

Let $\bar{k}\colon \coprod_{i,j} K_i \times L_{i,j} \longrightarrow \coprod_j K_j'$ be a bijection with $\bar{k}(\coprod_i K_i \times L_{i,j}) = K_j'$ for all j, and if $(k,\ell) \in K_i \times L_{i,j}$ with $k\ell \in \tilde{K}_{i,j}$ then $\bar{k}(k,\ell) = k\ell$.

We now apply Step A with δ and $(K_j')_j$ standing for ε and $(K_i)_i$ to get a partition of unity $(E_{j,k}')_{j\in I', k\in K_j'}$ in M_ω such that

$$(9) \quad \sum_j |K_j'|^{-1} \sum_{k,\ell} |\alpha_{k\ell^{-1}}(E_{j,\ell}') - E_{j,k}'|_\tau \leq 16\delta^{\frac{1}{2}}$$

$$(10) \quad \sum_j \sum_k |\beta_g(E_{j,k}') - E_{j,k}'|_\tau \leq 34\delta^{\frac{1}{2}}, \quad g \in A$$

and, moreover, analogues of the commutativity relations (6) and (7) hold.

From the (K_j') indexed partition of unity $(E_{j,m}')$ we obtain a (K_i)

indexed one $(E_{i,k})$ by letting for $i \in I$ and $k \in K_i$

$$E_{i,k} = \sum_j \sum_\ell E'_{j,m}$$

where $j \in J$, $\ell \in L_{i,j}$ and $m = \bar{k}(k,\ell)$.

For $g \in A$ we have from (10)

$$\sum_{i,k} |\beta_g(E_{i,k}) - E_{i,k}|_\tau \leq 34\delta^{\frac{1}{2}} \quad . \tag{11}$$

Let $i \in I$ and $k_1, k_2 \in K_i$. We infer

$$\alpha_{k_1 k_2^{-1}}(E_{i,k_2}) - E_{i,k_1} = \sum_j |K'_j|^{-1} \sum_{\ell,k'} \alpha_{k_1 \ell k_1'^{-1}}(\alpha_{k' k'^{-1}}(E'_{j,k'}) - E'_{j,k_1'})$$

$$- \sum_j |K'_j|^{-1} \sum_{\ell,k'} \alpha_{k_1 \ell k_2'^{-1}}(\alpha_{k_2' k'^{-1}}(E'_{j,k'}) - E'_{j,k_2'})$$

$$+ \sum_j |K'_j|^{-1} \sum_{\ell,k'} (\alpha_{k_1 \ell k_1'^{-1}}(E'_{j,k'}) - E'_{j,k'})$$

$$- \sum_j |K'_j|^{-1} \sum_{\ell,k'} \alpha_{k_1 k_2^{-1}}(\alpha_{k_2 \ell k_2'^{-1}}(E'_{j,k_2'}) - E'_{j,k_2'})$$

where $j \in J$, $\ell \in L_{i,j}$, $k' \in K'_j$, $k_1' = \bar{k}(k_1,\ell)$, $k_2' = \bar{k}(k_2,\ell)$. Summing up we get

$$\sum_i |K_i|^{-1} \sum_{k_1,k_2} |\alpha_{k_1 k_2^{-1}}(E_{i,k_2}) - E_{i,k_1}|_\tau \leq 2\Sigma_1 + 2\Sigma_2$$

where

$$\Sigma_1 = \sum_j |K'|^{-1} \sum_{k_1',k'} |\alpha_{k_1' k'^{-}}(E'_{j,k'}) - E'_{j,k_1'}|_\tau$$

with $j \in J$, $k_1', k' \in K_j$, and

$$\Sigma_2 = \sum_i \sum_j \sum_{k,\ell} |\alpha_{k\ell k'^{-1}}(E'_{j,k'}) - E'_{j,k'}|_\tau$$

where $i \in I$, $k \in K_i$, $j \in J$, $\ell \in L_{i,j}$ and $k' = \bar{k}(k,\ell)$. We have from (9) $\Sigma_1 \leq 16\delta^{\frac{1}{2}}$.

On the other hand, from the definition (8) of $\tilde{K}_{i,j}$, we remark that if in Σ_2 we have $k' \in \tilde{K}_{i,j}$, then $k\ell = k'$ and the corresponding term in Σ_2 vanishes. Hence

$$\Sigma_2 \leq 2 \sum_i \sum_{k'} |E'_{j,k'}|_\tau$$

where $j \in J$ and $k' \in K'_j \setminus (\bigcup_i \tilde{K}'_{i,j})$.

We have for each $j \in J$, $|K'_j \setminus \bigcup_j \tilde{K}_{i,j}| \leq 4\varepsilon|K'_j|$ and hence the global estimates 6.1(5) with the constants corresponding to (9) above yield

$$\Sigma_2 \leq 2(4\varepsilon + 16\delta^{\frac{1}{2}}) \leq 8\varepsilon + 32\delta^{\frac{1}{2}} \quad .$$

Hence

$$(12) \qquad \sum_i \sum_{k\ell} |\alpha_{k\ell^{-1}}(E_{i,\ell}) - E_{i,k}|_\tau \leq 2\Sigma_1 + 2\Sigma_2 \leq 16\varepsilon + 96\delta^{\frac{1}{2}} \quad .$$

Given any $\delta > 0$, and any finite $\bar{A} \subset \bar{G}$, there exists a partition of unity $(E_{i,k})_{i\in I,\, k\in K_i}$ in M_ω such that (11), (12) and also (3) and (4) above hold.

We may now apply the Index Selection Trick the same way we did in **6.3**, Step D, so as to make in (11) and (12) $\delta = 0$ and $\bar{A} = \bar{G}$, and thus obtain (1) and (2). The whole construction above could have been done in the relative commutant of any given countable subset of M_ω. The theorem is proved.

Chapter 7: COHOMOLOGY VANISHING

In what follows we study the low dimensional unitary valued cohomology for an action α of an amenable group G on a von Neumann algebra M. We show that if α is centrally free the one- and two-dimensional cohomology vanishes for the action induced on the centralizing algebra, and in the two-dimensional case obtain bounds on the solution in terms of the cocycle. The main result is that if α is centrally free, then the 2-cohomology vanishes on M itself (Theorem 1.1).

7.1 Let us begin with some technical preliminaries. The result that follows was proved in [4, Prop. 1.1.3] for M_ω, but the proofs remain valid for M^ω too.

PROPOSITION (A.Connes). *Let* M *be a* W^**-algebra with separable predual and* $\omega \in \beta\mathbb{N}\setminus\mathbb{N}$.

(1) *Any projection in* M^ω *has a representing sequence consisting of projections in* M.

(2) *Any partition of unity in projections in* M^ω *can be represented by a sequence of partitions of unity in projections in* M.

(3) *Let* v *be a partial isometry in* M^ω *with* $v^*v = e$, $vv^* = f$,

and let $(e^\nu)_\nu$, $(f^\nu)_\nu$ *be representing sequences for* e *and* f, *consisting of projections in* M *such that* $e^\nu \sim f^\nu$ *for all* ν, *then there exists a representing sequence* $(v^\nu)_\nu$ *for* v *such that* $v^{\nu *}v^\nu = e^\nu$ *and* $v^\nu v^{\nu *} = f^\nu$.

(4) *Any unitary in* M^ω *has a representing sequence consisting of unitaries in* M.

(5) *Any system of matrix units in* M^ω *can be represented by a sequence of matrix units in* M.

The rest of this section deals with several inequalities extending to infinite factor properties of the trace norms. Let ϕ be a faithful normal state on the W^*-algebra M, and $\omega \in \beta\mathbb{N}\setminus\mathbb{N}$. We define for $x \in M^\omega$, $|x|_\phi = \phi^\omega(|x|)$. This is not necessarily a norm, not being subadditive, but its restriction to M_ω is a trace norm. More generally the following result holds.

LEMMA. *For any* $x_1,\dots,x_n \in M^\omega$ *and* $y_1,\dots,y_n \in M_\omega$, *we have*

$$(6)\qquad \Big|\sum_{i=1}^n x_iy_i\Big|_\phi \le \sum_{i=1}^n \|x_i\|\,|y_i|_\phi\ .$$

Proof. For any $a_i,b_i \in M$, $i=1,\dots,n$, consider the polar decompositions

$$b_i = v_i|b_i|\ ,\qquad \sum_i a_ib_i = u\Big|\sum_i a_ib_i\Big|\ .$$

We infer

$$\begin{aligned}\phi\Big(\Big|\sum_i a_ib_i\Big|\Big) &= \sum_i \phi(u^*a_iv_i|b_i|)\\ &\le \sum_i \Big|\phi(|b_i|^{\frac12}u^*a_iv_i|b_i|^{\frac12})\Big| + \sum_i \|a_i\|\,\|b_i\|^{\frac12}\,\|[\phi,|b_i|^{\frac12}]\|\\ &\le \sum_i \|a_i\|\phi(|b_i|) + \sum_i \|a_i\|\,\|b_i\|^{\frac12}\,\|[\phi,|b_i|^{\frac12}]\|\ .\end{aligned}$$

If we apply this to representing sequences for x_i,y_i we obtain (6).

This result is very useful for estimates concerning partitions of unity $y_1,\dots,y_n$ in M_ω. Further on we work with the norms $\|x\|_\phi^\# = (\frac12\phi^\omega(x^*x+xx^*))^{\frac12}$, $x \in M^\omega$, connected to the preceding ones by means of the inequalities

$$(7)\qquad \|x\|_\phi^\# \le \big(\tfrac12(|x|_\phi + |x^*|_\phi)\,\|x\|\big)^{\frac12}$$

$$(8)\qquad |x|_\phi \le (2|e|_\phi)^{\frac12}\,\|x\|_\phi^\# \le 2^{\frac12}\,\|x\|_\phi^\#$$

where e is the left support of x.

Although $\|\cdot\|_\phi^\#$ is not unitarily invariant, it satisfies the following inequality:

$$(9) \quad \|uv-1\|_\phi^\# \leq 2^{\frac{1}{2}}(\|u-1\|_\phi^\# + \|v-1\|_\phi^\#)$$

for any unitaries $u,v \in M^\omega$. This is immediate from the identity

$$\|uv-1\|_\phi^{\#2} + \|vu-1\|_\phi^{\#2} = 2\|u-v^*\|_\phi^{\#2} = 4 - uv - vu - u^*v^* - v^*u^*$$

together with the inequality

$$\|u-v^*\|_\phi^\# \leq \|u-1\|_\phi^\# + \|v-1\|_\phi^\# \quad .$$

This yields inductively estimates for longer products of unitaries as well; we shall use for instance the fact that for $u_1,u_2,u_3,u_4 \in \mathcal{U}(M^\omega)$,

$$(10) \quad \|u_1u_2u_3u_4 - 1\|_\phi^\# \leq 2\sum_{i=1}^{4} \|u_i - 1\|_\phi^\# \quad .$$

7.2 In what follows G will be a discrete group, always assumed countable, and M will be a von Neumann algebra with separable predual. Recall that a 1-cocycle for α is a map $u: G \to \mathcal{U}(M)$ with $u_1 = 1$ and such that its coboundary ∂u is trivial, i.e.

$$(\partial u)_{g,h} = u_g\alpha_g(u_h)u_{gh}^* = 1 \quad , \qquad g,h \in G \; .$$

The perturbation of (u_g) by $v \in \mathcal{U}(M)$ is the cocycle $(\tilde{u}_g)$ with

$$\tilde{u}_g = vu_g\alpha_g(v^*) \; , \qquad g \in G$$

and we call (u_g) the coboundary of v if $(\tilde{u}_g) \equiv 1$.

PROPOSITION. *Let G be a discrete amenable group, let M be a von Neumann algebra with separable predual, and let (α_g) be an action of G on M_ω, strongly free (see **5.6**) and semiliftable. Assume that there exists a faithful normal state ϕ on M such that $\phi|Z(M)$ is preserved by $\alpha|Z(M)$. Then any cocycle $(v_g) \subset M_\omega$ for (α_g) is a coboundary. Moreover if N is any given countable subset of M_ω, which commutes with (v_g), then $v = \partial w$ with w in the relative commutant of N in M_ω.*

Proof. To give the idea of the proof suppose first that α would contain a copy of the left regular action $\mathrm{Ad}\,\lambda: G \to \mathrm{Aut}(\ell^\infty(G))$, commuting with (v_g), i.e. there would exist a partition of unity $(E_g)_{g\in G}$ in $\{v_g | g \in G\}' \cap M_\omega$ such that $\alpha_g(E_h) = E_{gh}$, $g,h \in G$. Then we could define $w = \sum_g v_g^* E_g$ and thus we would get a unitary satisfying

$$w^* \alpha_g(w) = \sum_{k,h} v_k E_k \alpha_g(v_h^*) E_{gh} = \sum_h v_{gh} \alpha_g(v_h^*) E_{gh} = v_g \quad .$$

This is a form of the Shapiro lemma in cohomological algebra.

In our actual framework, the Rohlin theorem is an approximate form of the left regular action containment, and analogous formulae give an approximate vanishing on the cohomology in M_ω. By means of the Index Selection Trick we eventually obtain exact vanishing in M_ω.

Let us begin the proof. Let $0 < \varepsilon < 1$ and let a finite subset F of G be given. Let $(K_i)_{i \in I}$ be an ε-paving family of subsets of G which are (ε, F) invariant. We are under the hypothesis of the Rohlin theorem 6.1 and so we can find a partition of unity $(E_{i,k})_{i \in I,\, k \in K_i}$ in M_ω such that for any $i, j \in I$, $k, \ell \in K_i$, $m \in K_j$, $g, h \in G$ we have

$$\sum_i |K_i|^{-1} \sum_{k,\ell} |\alpha_{k\ell^{-1}}(E_{i,\ell}) - E_{i,k}|_\phi \leq 5\varepsilon^{\frac{1}{2}}$$

$$\alpha_g \alpha_h(E_{i,k}) = \alpha_{gh}(E_{i,k})$$

$$[\alpha_g(E_{i,k}), E_{j,m}] = 0$$

$$[\alpha_g(E_{i,k}), v_h] = 0 \quad .$$

We define the unitary $w \in M_\omega$ by

$$w = \sum_i \sum_k v_k^* E_{i,k} \qquad i \in I, \quad k \in K_i \quad .$$

Let $\tilde{v}_g = w v_g \alpha_g(w^*)$ be the perturbed cocycle. Let us keep $g \in F$ fixed. We infer

$$\tilde{v}_g - 1 = \sum_{i,j} \sum_{k,\ell} (v_k^* v_g \alpha_g(v_\ell) - 1) E_{i,k} \alpha_g(E_{j,\ell})$$

$$= \Sigma_1 + \Sigma_2 + \Sigma_3$$

where $i, j \in I$, $k \in K_i$, $\ell \in K_j$ and in Σ_1 we sum for $i = j$, $\ell \in K_i \cap g^{-1} K_i$, $k = g\ell$; in Σ_2 we sum for $i = j$, $\ell \in K_j \cap g^{-1} K_j$, $k \neq g$ and in Σ_3 for $\ell \in K_j \setminus g^{-1} K_j$.

From the cocycle identity we get $\Sigma_1 = 0$. Trace norm inequalities yield

$$|\Sigma_3|_\phi \leq 2 \sum_{j,\ell} |E_{j,\ell}|_\phi \qquad j \in I, \quad \ell \in K_j \setminus g^{-1} K_j \quad .$$

Since we have assumed $|K_i \setminus g^{-1} K_i| \leq \varepsilon |K_i|$ from the global estimates 6.1(5), we infer

(1) $\quad |\Sigma_3|_\phi \leq 2(\varepsilon + 5\varepsilon^{\frac{1}{2}}) \leq 12\varepsilon^{\frac{1}{2}} \quad .$

On the other hand,

$$|\Sigma_3|_\phi \leq 2 \sum_{i,j} \sum_{k,\ell} |E_{i,k}\alpha_g(E_{j,\ell})|_\phi$$

where $i,j \in I$; $k \in K_i$; $\ell \in K_j \cap g^{-1}K_j$ and either $i \neq j$ or $k \neq g\ell$. We obtain

$$(2) \quad |\Sigma_2|_\phi \leq 2 \sum_{j,\ell} |(1 - E_{j,g\ell})\alpha_g(E_{j,\ell})|_\phi$$

$$= 2 \sum_{j,\ell} |(1 - E_{j,g\ell})(E_{j,g\ell} - \alpha_g(E_{j,\ell}))|_\phi$$

$$\leq 2 \sum_{j,\ell} |E_{j,g\ell} - \alpha_g(E_{j,\ell})|_\phi$$

for $j \in I$ and $\ell \in K_j \cap g^{-1}K_j$. The estimates 6.1(4) yield

$$|\Sigma_2|_\phi \leq 2.10\,\varepsilon^{\frac{1}{2}} = 20\varepsilon^{\frac{1}{2}} \quad .$$

Summing up, we infer for $g \in F$,

$$|\tilde{v}_g - 1|_\phi \leq |\Sigma_1|_\phi + |\Sigma_2|_\phi + |\Sigma_3|_\phi \leq 32\varepsilon^{\frac{1}{2}}$$

Let us now take $\varepsilon = \frac{1}{n}$ and $F = F_n \subset\subset G$, where $F_n \nearrow G$; $n \in \mathbb{N}$. We obtain for each n a perturbation $w^{(n)}$ such that the corresponding perturbed cocycles $(\tilde{v}_g^{(n)})$ satisfy for any $g \in G$

$$\lim_{n\to\infty} |\tilde{v}_g^{(n)} - 1|_\phi = 0 \quad .$$

The Index Selection Trick, applied the same was as in the proof of Lemma 6.3, Step D, yields a unitary w such that the perturbed cocycle is trivial.

We could do the whole proof above in the relative commutant of a countable subset of M and thus obtain the supplementary assertion of the proposition. The proof is finished.

7.3 Let G again be a discrete countable group, and M a von Neumann algebra with separable predual. Recall that a cocycle crossed action $((\alpha_g),(u_{g,h}))$ of G on M is a pair of maps $\alpha\colon G \to \operatorname{Aut} M$ and $u\colon G \times G \to \mathcal{U}(M)$ such that $\alpha_1 = 1$,

$$\alpha_g\alpha_h = \operatorname{Ad} u_{g,h}\alpha_{gh} \qquad g,h \in G$$

and u is normalized by $u_{1,g} = u_{g,1} = 1$, $g \in G$ and satisfies

$$u_{g,h}u_{gh,k} = \alpha_g(u_{h,k})u_{g,hk} \qquad g,h,k \in G \quad .$$

A perturbation of $((\alpha_g),(u_{g,h}))$ is a family (v_g) of unitaries in M,

$g \in G$, with $v_1 = 1$; the corresponding perturbed cocycle crossed action $((\tilde{\alpha}_g),(\tilde{u}_{g,h}))$ is given by

$$\tilde{\alpha}_g = \operatorname{Ad} v_g \alpha_g$$

$$\tilde{u}_{g,h} = v_g \alpha_g(v_h) u_{g,h} v_{gh}^* \quad .$$

We omit the simple verification that this is a cocycle crossed action indeed. We say that $(u_{g,h})$ is the coboundary of (v_g) if $\tilde{u}_{g,h} \equiv 1$.

A simple but very useful remark is that the effect of two consecutive perturbations of (α,u), first with v and then with $\tilde{v}$, is the same as the one of perturbation with $\tilde{v}v$. Also, if v perturbs (α,u) to $(\tilde{\alpha},\tilde{u})$ and $u \equiv \tilde{u} \equiv 1$, then v is an α-cocycle.

We next show that we can perturb any cocycle $(u_{g,h})$ with some $(\bar{v}_g)$ to $(\bar{u}_{g,h})$ such that $(\bar{u}_{g,h})$ is approximately periodic in h with respect to the plaques of the Paving Structure, i.e. for any $p \in \mathbb{N}$, according to the approximate decomposition of the plaques, $K_j^{p+1} \simeq \bigcup_i \bigcup_\ell K_i^p \ell$, $i \in I_p$, $\ell \in L_{i,j}^p$, we have $\bar{u}_{g,h\ell} = \bar{u}_{g,h}$ for most $h \in K_i^p$ and $\ell \in L_{i,j}^p$. Moreover, $\bar{v}_g - 1$ is kept under control. This way if $u_{g,h} - 1$ is small for $h \in \bigcup_i K_i^n$, then $\bar{u}_{g,h} - 1$ is small for most $h \in G$. We use the notation in **3.4** for the Paving Structure.

LEMMA (Almost Periodization Lemma). *Let $((\alpha_g),(u_{g,h}))$ be a cocycle crossed action of the amenable group G on the von Neumann algebra M. Assume that a choice of a Paving Structure is made for G and use the notations in 3.4 for its elements. Then there exists a perturbation $(\bar{v}_g)$ of $((\alpha_g),(u_{g,h}))$ such that the perturbed cocycle crossed action $((\bar{\alpha}_g),(\bar{u}_{g,h}))$ satisfies for any $n \geq 1$, $j \in I_{n+1}$ and $g \in G_n$*

$$(1) \quad \left|\{h \in K_j^{n+1} \mid \bar{u}_{g,h} \neq \bar{u}_{g,k} \text{ for } \bar{k}^n(h) = (k,\ell),\ \ell \in \bigcup_i L_{i,j}^n\}\right| \leq 6\varepsilon_n |K_j^{n+1}|$$

Moreover, $(\bar{v}_g)$ and $(\bar{u}_{g,h})$ have the following property. If for some $n \in \mathbb{N}$, $\delta > 0$, and normal state ϕ on M,

$$\|u_{g,h} - 1\|_\phi^\# \leq \delta \qquad g,h,gh \in G_{n+1}$$

then

$$\|\bar{v}_g - 1\|_\phi^\# \leq 8^n \delta \qquad g \in G_{n+1}$$

$$\|\bar{u}_{g,h} - 1\|_\phi^\# \leq 8^n \delta \qquad g \in G_n; \quad h,gh \in G_{n+1} \; .$$

<u>Proof</u>. Let $n \in \mathbb{N}$ and let $H_{n+1} = (\bigcup_{i,j} \bar{K}_{i,j}^{n+1}) \setminus (G_{n+1} \cup \bigcup_{i,j} L_{i,j}^n)$. This set is contained in $\bigcup_{i,j} \bar{K}_{i,j}^{n+1}$, the subset of $\bigcup_j K_j^{n+1}$ which behaves well with respect to the approximate decomposition in plaques

$K^{n+1} \simeq \bigcup_i K_i^n L_{i,j}^n$ (see **3.4**).

Let $g \in H_{n+1}$ and let $i \in I_n$, $j \in I_{n+1}$ with $g \in \bar{K}_{i,j}^{n+1}$. From the definition of $\bar{K}_{i,j}^{n+1}$, i and j are uniquely determined and there exist unique $(k,\ell) \in K_i^n \times L_{i,j}^n$ with $g = k\ell$. Let $((\alpha_g^1),(u_{g,h}^1)) = ((\alpha_g),(u_{g,h}))$ and define inductively the perturbations

$$(2) \qquad v_g^n = \begin{cases} u_{k,\ell}^n & \text{if } g \in H_{n+1} \text{ and } g = k\ell \text{ as above} \\ 1 & \text{for the other } g \in G \setminus H_{n+1} \end{cases}$$

and let $((\alpha_g^{n+1}),(u_{g,h}^{n+1}))$ be the cocycle crossed action obtained by perturbing $((\alpha_g^n),(u_{g,h}^n))$ with (v_g^n); do this successively for $n = 1,2,3,\ldots$. We shall show that $u_{g,h}^n$ is approximately periodic at the level n, that this property is not destroyed by the next perturbations, and that the product of the perturbations v_g^n is stationary for each $g \in G$.

<u>Step A</u>. We show that if $g \in H_{n+1}$, and if $i \in I_n$, $j \in I_{n+1}$ are such that $g = k\ell$ with $(k,\ell) \in K_i^n \times L_{i,j}^n$, then $u_{k,\ell}^{n+1} = 1$. Indeed, we have $k,\ell \in K_i^n \cup L_{i,j}^n \subseteq G \setminus H_{n+1}$ and so $v_k^n = v_\ell^n = 1$; since we have $v_g^n = u_{k,\ell}^n$ we infer $u_{k,\ell}^{n+1} = v_k^n \alpha_k^n(v_\ell^n) u_{k,\ell}^n v_k^{n*} = u_{k,\ell}^n v_g^{n*} = 1$.

<u>Step B</u>. We now prove the approximate periodicity. If $g \in G_n$, $h, gh \in H_{n+1}$ and $h = k\ell$ with $k \in K_i^n$, $\ell \in L_{i,j}^n$ and if moreover $gk \in K_i^n$, then since $gh = (gk)\ell$, from the Step A we have $u_{k,\ell}^{n+1} = u_{gk,\ell}^{n+1} = 1$. Since u is a cocycle,

$$u_{g,h}^{n+1} = u_{g,k\ell}^{n+1} = \alpha_g^{n+1}(u_{k,\ell}^{n+1})^* \, u_{g,k}^{n+1} u_{gk,\ell}^{n+1} = u_{g,k}^{n+1}$$

hence the approximate periodicity relation holds for (g,h). For given $j \in I_{n+1}$ and $g \in G_n$ we evaluate the cardinality of the subset Δ_j^{n+1} of K_j^{n+1}, consisting of those h for which (g,h) does not satisfy the conditions above. We have

$$\Delta_j^{n+1} \subseteq \{1,g^{-1}\}(K^{n+1} \setminus H_{n+1}) \cup (K_j^{n+1} \setminus g^{-1}K_j^{n+1}) \cup \left(\bigcup_i (K_i^n \setminus g^{-1}K_i^n)\, L_{i,j}^n\right) .$$

We have shown in **3.4** that $|\bar{K}_{i,j}^{n+1}| \geq (1-\varepsilon_n)|K_i^n|\,|L_{i,j}^n|$, hence

$$|K_j^{n+1} \setminus \bigcup_i \bar{K}_{i,j}^{n+1}| \leq \varepsilon_n \sum_i |K_i^n|\,|L_{i,j}^n| = \varepsilon_n |K_j^{n+1}| .$$

In **3.5** we have assumed that for each $j \in I_{n+1}$

$$|G_{n+1} \cup \bigcup_i L^n_{i,j}| \leqslant \varepsilon_n |K^{n+1}_j| ,$$

$$|K^{n+1}_j \setminus H_{n+1}| \leqslant 2\varepsilon_n |K^{n+1}_j| .$$

From the left invariance properties of K^n_i, K^{n+1}_j to $g \in G_n \subseteq G_{n+1}$ we have

$$|K^{n+1}_j \setminus g^{-1} K^{n+1}_j| \leqslant \varepsilon_{n+1} |K^{n+1}_j|$$

$$|K^n_i \setminus g^{-1} K^n_i| \leqslant \varepsilon_n |K^n_i|$$

so that

$$\sum_i |K^n_i \setminus g^{-1} K^n_i| \, |L^n_{i,j}| \leqslant \varepsilon_n \sum_i |K^n_i| \, |L^n_{i,j}| = \varepsilon_n |K^{n+1}_j|$$

and finally

$$|\Delta^{n+1}_j| \leqslant (2.2\varepsilon_n + \varepsilon_{n+1} + \varepsilon_n) |K^{n+1}_j| \leqslant 4\varepsilon_n |K^{n+1}_j| .$$

Since $K^{n+1}_j \subseteq G_{n+2} \subseteq G_{n+1} \dots$, for any $g \in G$ there is at most one n for which $v^n_g \neq 1$; hence the product

$$v_g = \dots v^n_g v^{n-1}_g \dots v_g$$

is well defined. Again by the assumptions of **2.5**, $G_n(\bigcup_j K^{n+1}_j) \subseteq G_{n+2}$, and so if $g \in G_n$ and $h \in \bigcup_j K^{n+1}_j$, then $g,h,gh \in G_{n+2}$ and $u^{n+p}_{g,h} = u^{n+1}_{g,h}$ for any $p \geqslant 1$.

Since $((\bar{\alpha}_g),(\bar{u}_{g,h}))$, which is the perturbed of $((\alpha_g),(u_{g,h}))$ by $(\bar{v}_g)$, is also equal to the pointwise limit of $((\alpha^n_g),(u^n_{g,h}))$ when $n \to \infty$, the conclusion (1) is proved.

Step C. We prove the estimates. Let $L = \bigcup_{p<n} \bigcup_{i,j} L^p_{i,j}$. We have assumed in **3.5** that $L \subseteq G_{n+1}$. Let us define

$$A = \{(g,h) \in G_n \times G_{n+1} | gh \in G_{n+1}\}$$

$$B = \{(g,h) \in G_{n+1} \times L | gh \in G_{n+1}\} .$$

We prove inductively for $p = 1,2,\dots,n+1$ that

$$(3,p) \qquad \|u^p_{g,h} - 1\|^\#_\phi \leqslant 8^{p-1}\delta \qquad (g,h) \in A \cup B$$

$$(4,p) \qquad \|v^p_g - 1\|^\#_\phi \leqslant 8^{p-1}\delta \qquad g \in G_{n+1} .$$

From the definition of v^p_g, (4,p) follows from (3,p). By the hypothesis, (3,1) is true, since $A \cup B \subseteq G_{n+1} \times G_{n+1}$. Suppose that (3,p) and (4,p) are true for some p, $1 \leqslant p \leqslant n$, and let us prove (3,p+1). Let $(g,h) \in A \cup B$.

Suppose first that $v_h^p = 1$. Then

$$u_{g,h}^{p+1} = v_g^p \, u_{g,h}^p \, v_{gh}^{p*}$$

and since $g, gh \in G_{n+1}$, we may use (3,p) and (4,p) to conclude with the inequality **7.1**(10)

$$\|u_{g,h}^{p+1} - 1\|_\phi^\# \leqslant 2(\|v_g^p - 1\|_\phi^\# + \|u_{g,h}^p - 1\|_\phi^\# + \|v_{gh}^p - 1\|_\phi^\#)$$
$$\leqslant 6.8^{p-1}\delta < 8^p\delta \ .$$

Now let $v_h^p \neq 1$. Since $(g,h) \in A \cup B$, we have $h \in G_{n+1}$. From the definition of v_h^p we infer $p < n$. Since $h \in (\bigcup_j K_j^{p+1}) \setminus (G_{p+1} \cup \bigcup_{i,j} L_{i,j}^p)$, we have $h \notin \bigcup_{q \le p} \bigcup_{i,j} L_{i,j}^q \subset G_{p+1}$. From the assumptions **3.5**,

$$(\bigcup_j K_j^{p+1}) \cap (\bigcup_{q=p+1} L_{i,j}^q) = \emptyset \ .$$

Hence $h \notin L$ and so $(g,h) \in A$. There exist $i \in I_n$, $j \in I_{n+1}$, $k \in K_i^p$, $\ell \in L_{i,j}^p$ such that $h = k\ell$. Then the cocycle identity yields

$$u_{g,h}^{p+1} = v_g^p \, \alpha_g^p(v_h^p) u_{g,h}^p v_{gh}^{p*} = v_g^p \alpha_g^p(u_{k,\ell}^p) u_{g,k\ell}^p \, v_{gh}^{p*}$$
$$= v_g^p u_{g,k}^p \, u_{gk,\ell}^p \, v_{gh}^{p*} \ .$$

We again use assumptions **3.5** on the Paving Structure. We have $g, gh \in G_{n+1}$ since $(g,h) \in A$. Since $k \in K_i^p \subseteq G_{n+1}$ and $gk \in G_n K_i^p \subseteq G_{n+1}$ we infer $(g,k) \in A$. As $\ell \in L_{i,j}^p \subseteq L$, $gk \in G_{n+1}$ and $gk\ell = gh \in G_{n+1}$ we have $(gk,\ell) \in B$. The induction hypothesis yields

$$\|u_{g,h}^{p+1} - 1\|_\phi^\# \leqslant 2(\|v_g^p - 1\|_\phi^\# + \|v_{gh}^p - 1\|_\phi^\# + \|u_{g,k}^p - 1\|_\phi^\# + \|u_{gk,\ell}^p - 1\|_\phi^\#)$$
$$\leqslant 2 \times 4 \times 8^{p-1}\delta = 8^p\delta$$

and thus we have finished the proof of (3,p+1). Hence (3,p) and (4,p) hold for all $1 \leqslant p \leqslant n+1$. We have shown in Step B that for $g \in G_{n+1}$ we have $\bar{v}_g = v_g^p$ for some $p \leqslant n+1$, and for $g,h,gh \in G_n$ we have $\bar{u}_{g,h} = u_{g,h}^{n+1}$. The estimates in the conclusion of the lemma are thus proved.

<u>Remark</u>. If ϕ is a trace on M, we may work in Step C with $|\cdot|_\phi$ instead of $\|\cdot\|_\phi^\#$ and use a trace norm equality instead of **7.1**(10) to prove the following assertion.

If for some $\delta > 0$ and $n > 1$,

$$|u_{g,h} - 1|_\phi \leq \delta \qquad g,h,gh \in G_{n+1}$$

then

$$|\bar{v}_g - 1|_\phi \leq 4^n\delta \qquad g \in G_{n+1}$$

$$|\bar{u}_{g,h} - 1|_\phi \leq 4^n\delta \qquad g \in G_n\,, \quad h,gh \in G_{n+1}\,.$$

7.4 We now prove a vanishing result for M_ω-valued 2-cohomology; by means of the Almost Periodization Lemma we are able to obtain bounds for the solution.

PROPOSITION. *Let* G *be a discrete countable group, let* M *be a von Neumann algebra with separable predual and let* $((\alpha_g),(u_{g,h}))$ *be a cocycle crossed action of* G *on* M_ω, *semiliftable and strongly free. Let* ϕ *be a faithful normal state on* M, *such that* $\phi|Z(M)$ *is fixed by* $\alpha|Z(M)$. *Then* $(u_{g,h})$ *is a coboundary. Given* $n \in \mathbb{N}$, $n \geq 2$, *if*

$$|u_{g,h} - 1|_\phi \leq \varepsilon_{n-2} \quad \textit{for} \quad g \in G_n\,, \quad h,gh \in G_{n+1}$$

then $u = v$ *with*

$$|v_g - 1|_\phi \leq 18\varepsilon_{n-2} \quad \textit{for} \quad g \in G_{n-2}\,.$$

where $\varepsilon_n > 0$ *and* $G_n \subset\subset G$ *were defined in the Paving Structure* 3.4. *If moreover* $(u_{g,h}) \subset N' \cap M_\omega$ *for some countable* $N \subset M_\omega$, *we may take* $(v_g) \subset N' \cap M_\omega$ *as well.*

The proof will be done by successively perturbing $(u_{g,h})$ with a sequence of perturbations; its product converges such that at the limit we obtain the identity cocycle.

The lemma that follows displays the result of an application of the Rohlin lemma - Shapiro lemma, followed by the Approximate Periodicity lemma, and provides the inductive step in the proof of the proposition.

LEMMA. *By the conditions of the proposition, let* $n \geq 2$ *and suppose that the cocycle crossed action* $((\alpha_g),(u_{g,h}))$ *satisfies the following condition:*

For any $g \in G_{n-2}$ *and* $j \in I_n$ *there exists a set* $\Delta_j^n(g) \subset K_j^n$, *such that* $|\Delta^n(g)| \leq 7\varepsilon_{n-2}|K_j^n|$ *and for any* $g \in G_{n-2}$ *and* $h,gh \in \bigcup_j (K_j^n \setminus \Delta_j^n(g))$ *we have*

(1) $|u_{g,h} - 1|_\phi \leq \varepsilon_{n-2}\,.$

Then for each $g \in G_{n-1}$ *and* $j \in I_{n+1}$ *there is a set* $\Delta_j^{n+1}(g) \subset K_j^{n+1}$, *with* $|\Delta_j^{n+1}(g)| \leq 7\varepsilon_{n-1}|K_j^{n+1}|$ *and there exists a perturbation* (v_g) *of*

$((\alpha_g),(u_{g,h}))$ *such that the perturbed cocycle crossed action* $((\bar{\alpha}_g),(\bar{u}_{g,h}))$ *satisfies*

(2) $|\bar{u}_{g,h} - 1|_\phi \leqslant \varepsilon_{n-1}$

for $g \in G_{n-1}$; $h, gh \in G_n$ *and also for* $g \in G_{n-1}$, $h, gh \in \bigcup_j (K^{n+1} \setminus \Delta^{n+1}(g))$. *Moreover the perturbation satisfies*

(3) $|v_g - 1|_\phi \leqslant 18\varepsilon_{n-2}, \quad g \in G_{n-2}$.

Proof. Step A. We again use the Rohlin theorem and a form of Shapiro's lemma, the same way as for the 1-cohomology, to obtain the approximate vanishing of the cohomology.

Recall from the Paving Structure that $(K_i^n)_{i \in I_n}$ was an ε_n-paving family of sets and ℓ_g^n was the approximate left translation with $g \in G$ on $\bigcup_i K_i^n$.

We perturb $((\alpha_g),(u_{g,h}))$ by $(\tilde{v}_g)$ to $((\tilde{\alpha}_g),(\tilde{u}_{g,h}))$ such that

(4) $|\tilde{u}_{g,h} - 1|_\phi \leqslant 32\varepsilon_n^{1/2} \quad g, g, gh \in G_n$;

(5) $|\tilde{v}_g - 1|_\phi \leqslant 16\varepsilon_{n-2} \quad g \in G_{n-2}$.

According to the Rohlin theorem 6.1 let us choose a partition of unity $(E_{i,k})$, $i \in I_n$, $k \in K_i^n$ such that

$$\sum_i |K_i^n|^{-1} \sum_{k,\ell} |\alpha_{k\ell^{-1}}(E_{i,\ell}) - E_{i,k}|_\phi < 5\varepsilon_n^{1/2}$$

$$\alpha_g(\alpha_h(E_{i,k})) = \alpha_{gh}(E_{i,k})$$

$$[\alpha_g(E_{i,k}), E_{j,\ell}] = 0$$

$$[\alpha_g(E_{i,k}), u_{g,h}] = 0 \quad \text{for all} \quad i,j,k,\ell,h,g .$$

Let us define the unitary for $g \in G$,

$$\tilde{v}_g = \sum_{i,k} u_{g,k}^* E_{i,h}$$

where $i \in I_n$, $k \in K_i^n$ and $h = \ell_g^n(k)$. Let us keep $g, h \in G_n$ with $gh \in G_n$ fixed, and for $i \in I_n$ let

$$\tilde{K}_i^n = \{k \in K_i^n \mid hk \in K_i^n,\ ghk \in K_i^n\} .$$

Since K_i^n is (ε_n, G_n) invariant, $|\tilde{K}_i^n| \geqslant (1-\varepsilon_n)|K_i^n|$. We infer from the definition of $\tilde{v}_g$ and $\tilde{u}_{g,h}$

$$\begin{aligned}\tilde{u}_{g,h} - 1 &= \tilde{v}_g \alpha_g(\tilde{v}_h) u^*_{g,h} \tilde{v}_{gh} - 1 \\ &= \sum_{i,j} \sum_{k,\ell} (u^*_{g,k} \alpha_g(u^*_{h,\ell}) u_{g,h} u_{gh,m} - 1) E_{i,p} \alpha_g(E_{j,q}) \\ &= \Sigma_1 + \Sigma_2 + \Sigma_3\end{aligned}$$

where $i,j \in I_n$, $k \in K^n_i$, $\ell \in K^n_j$, $p = \ell^n_g(k)$, $q = \ell^n_h(\ell)$, $m = (\ell^n_{gh})^{-1}(p) \in K^n_i$; in Σ_1 we sum for $i=j$ and $\ell = m \in \tilde{K}^n_i$; in Σ_2 we sum for $m \in K^n_i \setminus \tilde{K}^n_i$, and in Σ_3 for the remaining indices.

In Σ_1 we have $i=j$, $\ell = m$, $k = q = hm$, and the cocycle identity yields $\Sigma_1 = 0$. We have

$$|\Sigma_2|_\phi \leq 2 |\sum_{i,m} E_{i,p}|_\phi$$

where $i \in I_n$, $m \in K^n_i \setminus \tilde{K}^n_i$ and $p = \ell^n_{gh}(m)$. Since $|K^n_i \setminus \tilde{K}^n_i| \leq \varepsilon_n |K^n_i|$, the estimates 6.1(5) yield

$$|\Sigma_2|_\phi \leq 2(5\varepsilon_n^{1/2} + \varepsilon_n) \leq 12\varepsilon_n^{1/2} .$$

For the third sum we infer

$$|\Sigma_3|_\phi \leq 2 \sum_{i,j} \sum_{p,q} |E_{i,p} \alpha_g(E_{j,q})|_\phi$$

where $i,j \in I_n$, $q \in K^n_j \cap g^{-1} K^n_j$, $p \in K^n_i$ and $(i,p) \neq (j,q)$. We have already estimated an analogous sum in 7.2(2). In the same way we get

$$|\Sigma_3|_\phi \leq 20\varepsilon_n^{1/2} .$$

We have thus obtained for $g,h,gh \in G_n$

$$|\tilde{u}_{g,h} - 1|_\phi \leq |\Sigma_1|_\phi + |\Sigma_2|_\phi + |\Sigma_3|_\phi \leq 32\varepsilon_n^{1/2}$$

and thus we have proved (4).

Let us evaluate now the perturbation. Let $g \in G_{n-2}$. We decompose

$$\tilde{v}_g - 1 = \sum_{v,k} (u^*_{g,k} - 1) E_{i,h} = \Sigma_1 + \Sigma_2$$

where $j \in I_n$, $k \in K^n_j$, $h = \ell^n_j(k)$. In Σ_1 we sum for $k \in K^n_j \setminus \Delta^n_j(g)$ and in Σ_2 for $k \in \Delta^n_j(g)$. We infer from the hypothesis (1) of the lemma

$$|\Sigma_1|_\phi \leq \varepsilon_{n-2} .$$

On the other hand,

$$|\Sigma_2|_\phi \leq 2 \sum_{j,h} |E_{j,h}|_\phi$$

where $j \in I_n$, $h \in \ell_h^n(\Delta_j^n(g))$. Since for each j, $|\ell_g^n(\Delta_j^n(g))| \leq 7\varepsilon_{n-2}|K_j^n|$ by hypothesis, the estimates 6.1(5) of the Rohlin theorem yield

$$|\Sigma_2|_\phi \leq 2(7\varepsilon_{n-2} + 5\varepsilon_n^{\frac{1}{2}}) \leq 15\varepsilon_{n-2}$$

and thus using assumptions 3.5 on $(\varepsilon_n)_n$ we obtain

$$|\tilde{v}_g - 1|_\phi \leq 15\varepsilon_{n-2} + 10\varepsilon_n^{\frac{1}{2}} \leq 16\varepsilon_{n-2} \quad , \qquad g \in G_{n-2} \,.$$

Step B. A problem is that we have obtained in (4) $|\tilde{u}_{g,h} - 1|_\phi$ small for $h \in G_n$, but in the statement of the lemma we need it small for h in a larger set, for induction reasons. The gap is filled by the Almost Periodization Lemma 7.3.

Let us apply it to $((\alpha_g),(u_{g,h}))$, to obtain $(\bar{v}_g)$ perturbing it to $((\bar{\alpha}_g),(\bar{u}_{g,h}))$. Using the estimates in Remark 7.3, we infer from (4) and (5) in Step A

$$|\bar{u}_{g,h} - 1|_\phi \leq 4^{n-1}.32\varepsilon_n^{\frac{1}{2}} \leq \varepsilon_{n-1} \,, \qquad g \in G_{n-1},\ h,gh \in G_n \,;$$

$$|\bar{v}_g - 1|_\phi \leq 4^{n-1}.32\varepsilon_n^{\frac{1}{2}} \leq \varepsilon_{n-1} \quad , \qquad g \in G_n \,.$$

We now use the almost periodicity. Let $g \in G_{n-1}$ and for $h \in K_j^{n+1}$ let $\bar{k}^n(j,h) = (j_1,h_1,k_1)$ and $\bar{k}^{n-1}(j_1,h_1) = (j_2,h_2,k_2)$, where $\bar{k}^n\colon \coprod_j K_j^{n+1} \longrightarrow \coprod_{i,j} K_i^n \times L_{i,j}^n$ is the approximate decomposition defined in the Paving Structure. Let

$$\Delta_j'(g) \;=\; \{h \in K_j^{n+1} \mid \bar{u}_{g,h} \neq u_{g,h_1}\}$$

$$\Delta_j''(g) \;=\; \{h \in K_j^{n+1} \mid \bar{u}_{g,h_1} \neq u_{g,h_2}\}$$

Since the cocycle $(\bar{u}_{g,h})$ satisfies the almost periodicity property 7.3(1), we infer

$$|\Delta_j'(g)| \;\leq\; 6\varepsilon_n|K_j^{n+1}|$$

$$|\Delta_j''(g)| \;\leq\; \sum_i (|L_{i,j}^n|\, 6\varepsilon_{n-1}|K_i^n|) \;=\; 6\varepsilon_{n-1}|K_j^{n+1}| \quad .$$

So if we take $\Delta_j^{n+1}(g) = \Delta_j'(g) + \Delta_j''(g)$, then $\Delta_j^{n+1}(g) \leq (6\varepsilon_{n-1} + 6\varepsilon_n)|K_j^{n+1}| \leq 7\varepsilon_{n-1}|K_j^{n+1}|$.

For $h \in K_j^{n+1} \setminus \Delta_j^{n+1}(g)$ we have, with the notation above, $\bar{u}_{g,h} = \bar{u}_{g,h_2}$, and $h_2 \in K_j^{n-1} \subseteq G_n$. Hence

$$|\bar{u}_{g,h} - 1|_\phi \leq \varepsilon_{n-1} \quad , \quad g \in G_{n-1},\quad h \in K_j^{n+1} \setminus \Delta_j^{n+1}(g)$$

and statement (2) in the lemma is proved.

Let us now prove (3). Let $v_g = \bar{v}_g \tilde{v}_g$, such that $((\bar{\alpha}_g),(\bar{u}_{g,h}))$ is the perturbed of $((\alpha_g),(u_{g,h}))$ by (v_g). From previous estimates we infer

$$|v_g - 1|_\phi \leq |\bar{v}_g - 1|_\phi + |\tilde{v}_g - 1|_\phi \leq 16\varepsilon_{n-2} + \varepsilon_{n-1} \leq 17\varepsilon_{n-2}, \quad g \in G_{n-2}$$

and the proof of the lemma is finished.

7.5 Let us now prove Proposition 7.4, by applying successively the preceding lemma for $n, n+1, n+2, \dots$.

Let $((\alpha_g^n),(u_{g,h}^n)) = ((\alpha_g),(u_{g,h}))$ and for $p \geq n$, suppose given $((\alpha_g^p),(u_{g,h}^p))$ which satisfies

$$(1,p) \qquad |u_{g,h}^p - 1|_\phi \leq \varepsilon_{p-2}$$

for $g \in G_{p-2}$, $h, gh \in \bigcup_j (K_j^p \setminus \Delta_j^p(g))$, with $\Delta_j^p(g) \subset K_j^p$ and $|\Delta_j^p(g)| \leq 6\varepsilon_{p-2}|K_j^p(g)|$, $g \in G_{p-2}$, $j \in I_p$.

Since $\bigcup_i K_i^n \subseteq G_{n+1}$, (1,n) is true. We use 7.5 to perturb $((\alpha_g^p),(u_{g,h}^p))$ with (v_g^p) to $((\alpha_g^{p+1}),(u_{g,h}^{p+1}))$, satisfying (1,p+1) and

$$(2,p) \qquad |u_{g,h}^{p+1} - 1|_\phi \leq \varepsilon_{p-2} \quad \text{for} \quad g \in G_{p-1}, \quad h, gh \in G_p$$

$$|v_g^p - 1|_\phi \leq 17\varepsilon_{p-2} \quad \text{for} \quad g \in G_{p-2} \ .$$

Let $v_g^{(p)} = v_g^p v_g^{p-1} \dots v_g^p$ for $p \geq n$. If $g \in G_{p-2}$, then

$$|v_g^{(p)} - v_g^{(p-1)}|_\phi = |(v_{g-1}^p)\, v_g^{(p-1)}|_\phi = |v_g^p - 1|_\phi \leq 17\varepsilon_{p-2} \ .$$

Hence for $m \geq p \geq n-1$ and $g \in G_{p-2}$,

$$|v_g^{(m)} - v_g^{(p)}|_\phi \leq \sum_{k=p+1}^{m} 17\varepsilon_{k-2} \leq 18\varepsilon_{p-1}$$

where $v_g^{(n-1)} = 1$, and the assumptions on $(\varepsilon_n)_n$ have been used. Thus the sequence $v_g^{(p)}$ converges *-strongly to a unitary $v_g \in M_\omega$ for any $g \in \bigcup_p G_{p-2} = G$; moreover

$$|v_g - 1|_\phi \leq 18\varepsilon_{n-2}, \qquad g \in G_{n-2} \ .$$

Since $((\alpha_g^p),(u_{g,h}^p))$ is the perturbed of $((\alpha_g),(u_{g,h}))$ by $(v_g^{(p-1)})$, in view of (2,p) we infer $u = \partial v$. This ends the proof of Proposition 7.4.

7.6 The same techniques which in the preceding sections yielded the vanishing with bounds of the 2-cohomology on M_ω, also give the vanishing of the 2-cohomology with bounds on M. Some additional complication is due to the absence of a trace on M.

Let us recall for convenience Theorem 1.1, in a form in which the Paving Structure appears explicitly in the estimates.

THEOREM. *Let* G *be a discrete amenable group, and let* $((\alpha_g),(u_{g,h}))$ *be a cocycle crossed action of* G *on* M *which is centrally free. Let* ϕ *be a faithful normal state on* M, *such that* $\phi|Z(M)$ *is kept fixed by* $\alpha|Z(M)$. *Then* $(u_{g,h})$ *is a coboundary.*

Moreover, given $n \geq 2$ *and a finite set* $W \subset \mathcal{U}(M)$, *if we have*

$$\|u_{g,h} - 1\|_\psi^\# \leq \varepsilon_{n-2}\,, \qquad g \in G_{n-2},\quad h, gh \in G_{n+1},\quad \psi \in \Phi$$

where $\Phi = \{\mathrm{Ad}\, w\phi \,|\, w \in W\}$, *then* $u = \partial v$ *with*

$$\|v_g - 1\|_\psi^\# \leq 2\varepsilon_{n-2}^{1/2} \quad \textit{for} \quad g \in G_{n-2}\,, \quad \psi \in \Phi \quad .$$

In the proof of this theorem, we use $|\cdot|_\phi$ and the inequality 7.1(6) for estimates in connection with the partitions of unity in M_ω yielded by the Rohlin lemma, and the norm $\|\cdot\|_\phi^\#$ for the rest. The only problem appears in connection with the estimates giving the convergence of infinite products of perturbations, since $\|\cdot\|_\phi^\#$ is not unitarily invariant. We use the inequality

$$(1) \qquad \|xv\|_\phi^\# \leq 2^{1/2}(\|x\|_\phi^\# + \|x\|_{\mathrm{Ad}\, v\phi}^\#)\,, \qquad x \in M\,,\quad v \in \mathcal{U}(M)$$

which is immediate from the identity

$$\|xv\|_\phi^{\#2} + \|x^*v\|_\phi^{\#2} = \|x\|_\phi^{\#2} + \|x\|_{\mathrm{Ad}\, v\phi}^{\#2}$$
$$= \tfrac{1}{2}\phi(x^*x + xx^* + v^*x^*xv + v^*xx^*v) \quad .$$

We thus have to use an ever larger family of norms at each step, and what allows us to do so is the fact that the estimates in the Rohlin theorem and the Shapiro lemma depend only on $\phi|Z(M) = (\mathrm{Ad}\, v\phi)|Z(M)$.

7.7 The inductive step of the proof of Theorem 7.6 is provided by the following lemma, which is an analogue of Lemma 7.4.

LEMMA. *Let* $G, M, ((\alpha_g),(u_{g,h})), \phi$ *be as in the theorem. Let* $n \geq 2$ *and let* $\Phi_n \subset \Psi_{n+1}$ *be finite sets of normal states on* M, *which on*

$Z(M)$ *coincide with* $\phi|Z(M)$. *Suppose that*

$$\|u_{g,h}-1\|_{\psi}^{\#} \leq \varepsilon_{n-2} \quad \textit{for} \quad g \in G_{n-2}; \quad g,gh \in \bigcup_j (K_j^n \setminus \Delta_j^n(g)); \quad \psi \in \Phi_n$$

where the sets $\Delta^n(g) \subset K_j^n$; $g \in G_{n-2}$; $j \in I_n$ *satisfy* $|\Delta_j^n(g)| \leq 7\varepsilon_{n-2}|K_j^n|$. *Then there exists a perturbation* (v_g) *of* $((\alpha_g),(u_{g,h}))$ *such that*

$$\|v_g-1\|_{\psi}^{\#} \leq 9\varepsilon_{n-2}^{1/2} \qquad g \in G_{n-2}, \quad \psi \in \Phi_n$$

and the perturbed cocycle $((\bar{\alpha}_g),(\bar{u}_{g,h}))$ *satisfies*

$$\|\bar{u}_g-1\|_{\psi}^{\#} \leq \varepsilon_{n-1} \qquad \psi \in \Phi_{n+1}$$

for $g \in G_{n-1}$; $h,gh \in G_n$ *and also for* $g \in G_{n-1}$; $h,gh \in \bigcup_j (K_j^{(n+1)} \setminus \Delta_j^{n+1}(g))$,

$$\Phi_{n+1} = \{\mathrm{Ad}\, v_g\psi \,|\, g \in G_{n-1}, \ \psi \in \Psi_{n+1}\}$$

and for $j \in I_{n+1}$, $g \in G_{n-1}$, *the sets* $\Delta_j^{n+1}(g) \subseteq K_j^{n+1}$ *satisfy* $|\Delta_j^{n+1}(g)| \leq 7\varepsilon_{n-1}|K_j^{n+1}|$.

Proof. The proof will parallel the one of Lemma 7.4.

Step A. Let $((\beta_g),(U_{g,h}))$, where $\beta_g = (\alpha_g)^{\omega} \in \mathrm{Aut}\, M^{\omega}$ be the cocycle crossed action induced by $((\alpha_g),(u_{g,h}))$ on M^{ω}. Since $\mathrm{Ad}\, U_{g,h}|M_{\omega} = \mathrm{id}$, $(\beta_g|M_{\omega})$ is an action, which by Lemma 5.6 is strongly free. The Rohlin theorem yields a partition of the unity $(E_{i,k})$, $i \in I_n$, $k \in K_i^n$, in M_{ω} such that

$$\sum_i |K_i^n|^{-1} \sum_{k,\ell} |\beta_{k\ell^{-1}}(E_{i,\ell}) - E_{i,k}|_{\phi} < 5\varepsilon^{1/2}$$

$$[\beta_g(E_{i,k}), E_{j,\ell}] = 0 \quad \text{for all} \quad i,j,k,\ell,g \ .$$

We define the perturbation $(\tilde{V}_g) \subset M^{\omega}$ by

$$\tilde{V}_g = \sum_{i,k} U_{g,k}^{*} E_{i,h}$$

where $i \in I_n$, $k \in K_i^n$ and $h = \ell_g^n(k)$.

Let $((\tilde{\beta}_g),(\tilde{U}_{g,h}))$ be the cocycle crossed action obtained by perturbing $((\beta_g),(U_{g,h}))$ with $(\tilde{V}_g)$. For further use we need estimates of $\mathrm{Ad}\,\tilde{V}_k(\tilde{U}_{g,h}-1)$. The estimates of $\tilde{u}_{g,h}-1$ in 7.4(2) were based merely on estimates of the Rohlin partition $(E_{i,k})$ and did not involve any estimates on the cocycle $(u_{g,h})$ which was perturbed. Since in our present context any $\tilde{V}_k$ commutes with any $E_{i,h}$, the same estimates work, letting the inequality 7.1(6) replace the trace norm inequality. In this way we infer

$$|\mathrm{Ad}\ \tilde{V}_k(U_{g,h}-1)|_{\psi} \leq 32\varepsilon_n^{1/2}$$

and similarly

$$|\mathrm{Ad}\ \tilde{V}_k^*(\tilde{U}_{g,h}^* - 1)|_\psi \leq 32\varepsilon_n^{\frac{1}{2}}$$

for $k \in G$, $g,h,gh \in G_n$, $\psi \in \Psi_{n+1}$, where we have also used the fact that for $\psi \in \Psi_{n+1}$, $\psi_\omega = \phi_\omega$, since $\psi|Z(M) = \phi|Z(m)$.

Via the inequality 7.1(7) this easily yields

$$(1) \quad \|\mathrm{Ad}\ \tilde{V}_k^*(\tilde{U}_{g,h} - 1)\|_\psi^\# \leq (\tfrac{1}{2}(32\varepsilon_n^{\frac{1}{2}} + 32\varepsilon_n^{\frac{1}{2}}).2)^{\frac{1}{2}} = 8\varepsilon_n^{\frac{1}{2}}$$

for $k \in G$, $g,h,gh \in G_n$, $\psi \in \Psi_{n+1}$. On the other hand, in the same way as in 7.4(5), we have

$$|V_g - 1|_\psi \leq 16\varepsilon_{n-2} \quad , \qquad |\tilde{V}_g^* - 1|_\psi \leq 16\varepsilon_{n-2}$$

and hence

$$(2) \quad \|\tilde{V}_g - 1\|_\psi^\# \leq (\tfrac{1}{2}(16\varepsilon_{n-2} + 16\varepsilon_{n-2}).2)^{\frac{1}{2}} \leq 6\varepsilon_{n-2}^{\frac{1}{2}}$$

for $g \in G_{n-2}$, $\psi \in \Phi_n$.

<u>Step B</u>. We apply the Almost Periodization Lemma to $((\tilde{\beta}_g),(\tilde{U}_{g,h}))$ and perturb it with $(\bar{V}_g)$ to get $((\bar{\beta}_g),(\bar{U}_{g,h}))$. The estimates in Lemma 7.3 yield from (1) above

$$(3) \quad \|\mathrm{Ad}\ \tilde{V}_k^*(\bar{V}_g - 1)\|_\psi^\# \leq 8^{n-1}.8\varepsilon_n^{\frac{1}{4}} = 8^n\,\varepsilon_n^{\frac{1}{4}} \;, \qquad k \in G,\ g \in G_n,\ \psi \in \Psi_{n+1}$$

and

$$(4) \quad \|\mathrm{Ad}\ \tilde{V}_k^*(\bar{U}_{g,h} - 1)\|_\psi^\# \leq 8^{n-1}.8\varepsilon_n^{\frac{1}{4}} = 8^n\,\varepsilon_n^{\frac{1}{4}}, \qquad k \in G,\ g \in G_{n-1},\ h,gh \in G_n,\ \psi \in \Psi_{n+1} .$$

The sets $\Delta_j^{n+1}(g)$ are defined the same way as in **7.4** and, as there, because of the almost periodicity of $\bar{U}_{g,h}$, inequality (4) above holds for $g \in G_{n-1}$; $g,gh \in \bigcup_j (K_j^{(n+1)} \setminus \Delta_j^{n+1}(g))$ as well.

Let $V_g = \bar{V}_g\tilde{V}_g$, $g \in G$. We infer from (1) and (2) above, by means of 7.1(9),

$$\|V_g - 1\|_\psi^\# \leq 2^{\frac{1}{2}}(\|\bar{V}_g - 1\|_\psi^\# + \|\tilde{V}_g - 1\|_\psi^\#)$$
$$\leq 2^{\frac{1}{2}}(6\varepsilon_{n-2}^{\frac{1}{2}} + 8^n\,\varepsilon_n^{\frac{1}{4}}) < 9\varepsilon_{n-2}^{\frac{1}{2}}$$

for $g \in G_{n-2}$ and $\psi \in \Phi_n \subseteq \Psi_{n+1}$, where again the assumptions on (ε_n) have been used.

On the other hand, the estimates (1) and (2) yield, with 7.1(10),

$$(5) \quad \|\mathrm{Ad}\ V_k^*(\bar{U}_{g,h} - 1)\|_\psi^\# = \|(V_k^*\bar{V}_kV_k)(\mathrm{Ad}\ V_k^*(\bar{U}_{g,h}))(V_k^*\bar{V}_kV_k) - 1\|_\psi^\#$$
$$\leq 2(2\|\tilde{V}_k^*\bar{V}_k\tilde{V}_k - 1\|_\psi^\# + \|\mathrm{Ad}\ \tilde{V}_k^*(\bar{U}_{g,h} - 1)\|_\psi^\#)$$
$$\leq 2(2 \cdot 8^n\,\varepsilon_n^{\frac{1}{4}} + 8^n\,\varepsilon_n^{\frac{1}{4}}) < \varepsilon_{n-1}$$

for $k \in G_{n-2}$, $\psi \in \Psi_{n+1}$ and either $g \in G_{n-1}$, $h,gh \in G_n$ or $g \in G_{n-1}$, $h,gh \in \bigcup_j (K_j^{n+1} \setminus \Delta_j^{n+1}(g))$.

Let $(v_g^\nu)_\nu$ be representing sequences for V_g, with v_g^ν unitaries in M, $v_1^\nu = 1$, $\nu \in \mathbb{N}$. Let $((\bar{\alpha}_g),(\bar{u}_{g,h}))$ be the perturbed of $((\alpha_g),(u_{g,h}))$ by (v_g^ν). Then $(\bar{u}_{g,h}^\nu)_\nu$ represents $\bar{U}_{g,h}$, and so we may choose $\nu \in \mathbb{N}$ such that if $v_g = v_g^\nu$, $\bar{\alpha}_g = \bar{\nu}_g$ and $\bar{u}_{g,h} = \bar{u}_{g,h}^\nu$, then

$$\| v_g - 1 \|_\psi^\# \leqslant 9\varepsilon_{n-2}^{\frac{1}{2}} \qquad g \in G_{n-1}, \quad \psi \in \Phi_n$$

and also

$$\| \mathrm{Ad}\, v_k^*(\bar{u}_{g,h} - 1) \|_\psi^\# \leqslant \varepsilon_{n-1}, \quad k \in G_{n-1}, \ \psi \in \Psi_{n+1}$$

where either $g \in G_{n-1}$, $h,gh \in G_n$ or $g \in G_{n-1}$, $h,gh \in \bigcup_j (K_j^{n+1} \setminus \Delta_j^{n+1}(g))$. If $\psi \in \Phi_{n+1}$, then $\psi = \mathrm{Ad}\, v_k \bar{\psi}$ for some $k \in G_{n-1}$ and $\bar{\psi} \in \Psi_{n+1}$, and so

$$\| \bar{u}_{g,h} - 1 \|_\psi^\# = \| \mathrm{Ad}\, v_k^*(\bar{u}_{g,h} - 1) \|_{\bar{\psi}}^\# \leqslant \varepsilon_{n-1}$$

for g,h as before. The lemma is proved.

7.8 Let us now prove Theorem 7.6. We successively perturb the given cocycle with perturbations given by Lemma 7.8 for $n, n+1, n+2, \ldots$, as in the proof of **7.5**. Let $((\alpha_g^n),(u_{g,h}^n)) = ((\alpha_g),(u_{g,h}))$ and $\Phi_n = \Phi$. Suppose for $p \geqslant n$ that we are given for $k = n,\ldots,p$ a centrally free cocycle crossed action $((\alpha_g^k),(u_{g,h}^k))$, a finite set Φ_k of faithful normal states on M and a perturbation (v_g^k) of $((\alpha_g^k),(u_{g,h}^k))$ taking it into $((\alpha_g^{k+1}),(u_{g,h}^{k+1}))$, $k = n,\ldots,p-1$, such that

$$(1,p) \quad \| u_{g,h}^p - 1 \|_\psi^\# \leqslant \varepsilon_{p-2} \quad \text{for} \quad \psi \in \Phi_p,\ g \in G_{p-2};\ h,gh \in \bigcup_j (K_j^p \setminus \Delta_j^p(g))$$

where for $g \in G_{p-2}$, $j \in I_p$, we have $\Delta_j^p(g) \subseteq K_j^p$ and $|\Delta_j^p(g)| \leqslant 6\varepsilon_{p-2}|K_j^p|$.

For $p = n$, $(1,n)$ holds by hypothesis since $\bigcup_j K_j^n \subseteq G_{n+1}$.

We let $(v_g^{(n-1)}) \equiv 1$ and for $n \leqslant k < p$ we take $v_g^{(k)} = v_g^k v_g^{k-1} \ldots v_g^n$. We apply the previous lemma to $((\alpha_g^p),(u_{g,h}^p))$ with n replaced by p, Φ_p defined inductively above, and $\Psi_{p+1} = \{\mathrm{Ad}\, v_g^{(p-1)}\psi \mid g \in G_{p-2},\ \psi \in \Phi_n\}$. We obtain a perturbation (v_g^p) such that if $((\alpha_g^{p+1}),(u_g^{p+1}))$ denotes the cocycle crossed action $((\alpha_g^p),(u_{g,h}^p))$ perturbed by (v_g^p), if $v_g^{(p)} = v_g^p v_g^{(p-1)}$ and if $\Phi_{p+1} = \{\mathrm{Ad}\, v_g^p \psi \mid \psi \in \Psi_{p+1},\ g \in G_{p-1}\}$, then $(u_{g,h}^{p+1})$ satisfies the condition $(1,p+1)$ above, and also

$$(2,p) \quad \| u_{g,h}^{p+1} - 1 \|_\psi^\# \leqslant \varepsilon_{p-1} \quad \text{for } g \in G_{p-1},\ h,gh \in G_p \text{ and } \psi \in \Phi_{p+1}$$

and

$$\|v_g^p - 1\|_\psi^\# \leq 9\varepsilon_{p-2} \quad , \qquad g \in G_{p-2}\,, \quad \psi \in \Phi_p \quad .$$

Using the inequality 7.6(1), we infer for $g \in G_{p-2}$ and $\psi \in \Phi_n$,

$$\|v_g^{(p)} - v_g^{(p-1)}\|_\psi^\# = \|(v_g^p - 1)\, v_g^{(p-1)}\|_\psi^\# \leq 2^{\frac{1}{2}}(\|v_g^p - 1\|_\psi^\# + \|v_g^p - 1\|_{\psi_g}^\#)$$

where $\psi_g = \mathrm{Ad}\, v_g^{(p-1)}\psi$. But if $p > n$,

$$\psi_g = \mathrm{Ad}\, v_g^{p-1}(\mathrm{Ad}\, v_g^{(p-2)}\psi) \in \mathrm{Ad}\, v_g^{p-1}(\Psi_p) \subseteq \Phi_p$$

and so

$$\|v_g^{(p)} - v_g^{(p-1)}\|_\psi^\# \leq 2^{\frac{1}{2}}(9\varepsilon_{p-2}^{\frac{1}{2}} + 9\varepsilon_{p-2}^{\frac{1}{2}}) \leq 26\varepsilon_{p-2}^{\frac{1}{2}} \quad .$$

Hence for $m > p \geq n$, $\psi \in \Phi$ and $g \in G_{p-2}$ we have

$$\|v_g^{(m)} - v_g^{(p)}\|_\psi^\# \leq \sum_{k=p+1}^{m} 26\varepsilon_{k-2}^{\frac{1}{2}} \leq 27\varepsilon_{p-1}^{\frac{1}{2}} \quad .$$

Since $\varepsilon_p \searrow 0$ and $G_p \nearrow G$, the *-strong limit $v_g = \lim_p v_g^{(p)}$ exists for each $g \in G$ and satisfies for $g \in G_{n-2}$

$$\|v_g - 1\|_\psi^\# \leq 27\varepsilon_{n-2}\,, \qquad g \in G_{n-2}\,, \quad \psi \in \Phi_n$$

and since $((\alpha_g^p),(u_{g,h}^p))$ is the perturbed of $((\alpha_g),(u_{g,h}))$ by $(v_g^{(p-1)})$, and from (2,p) above, $\lim_{p\to\infty} u_{g,h}^p = 1$ *-strongly, $g,h \in G$, we infer $u = \partial v$. The theorem is proved.

Chapter 8: MODEL ACTION SPLITTING

In this chapter we prove Theorems 1.2 and 1.3, which assert that a centrally free action of an amenable group "contains", if perturbed by an arbitrarily close to 1 cocycle, both the trivial action and the model action. The proofs also yield the analogous results, Theorems 1.5 and 1.6, for G-kernels.

8.1 We begin with some technical lemmas. The first result is due to Connes ([4, Lemma 1.1.4]). The statement here is slightly stronger but follows from the same proof.

LEMMA 1. *Let* M *be a countably decomposable* W^**-algebra and let* Ψ *be a finite set of normal states of* M. *If* $e, f \in \text{Proj } M$ *and* $e \sim f$ *then there exists a partial isometry* $v \in M$ *with* $v^*v = e$, $vv^* = f$

$$\|v-f\|_{\psi}^{\#} \leq 6\|e-f\|_{\psi}^{\#}$$
$$\|v^*-f\|_{\psi}^{\#} \leq 7\|e-f\|_{\psi}^{\#}$$

for any $\psi \in \Psi$.

A similar result holds for the L^1-norm.

LEMMA 2. *Let* M *be a finite* W^**-algebra with a normal trace* τ. *If* $e, f \in \text{Proj } M$ *with* $e \sim f$ *then there exists a partial isometry* $v \in M$ *with* $v^*v = e$, $vv^* = f$ *and*

$$|v-f|_{\tau} \leq 3|e-f|_{\tau} .$$

Proof. Let $\varepsilon = |e-f|_{\tau}$. Let $fe = w\rho$ be the polar decomposition of fe and let $e_1 = w^*w \leq e$, $f_1 = ww^* \leq f$. We have

$$|w-f|_{\tau} \leq |w-fe|_{\tau} + |fe-f|_{\tau} = |w(e-\rho)|_{\tau} + |f(e-f)|_{\tau}$$
$$\leq |e-\rho|_{\tau} + |e-f|_{\tau} = |e-\rho|_{\tau} + \varepsilon .$$

Since $\rho^2 = efe \leq e$,

$$|e-\rho|_{\tau} \leq |e-\rho^2|_{\tau} = |e(e-f)e|_{\tau} \leq |e-f|_{\tau} = \varepsilon$$

hence $|w-f|_{\tau} \leq 2\varepsilon$.

Since M is finite, $f-f_1 \sim e-e_1$. Let us choose $u \in M$ with $u^*u = e-e_1$ and $uu^* = f-f_1$, and let $v = u+w$. Then $v^*v = e$ and $vv^* = f$. As $\rho^2 \leq e_1 \leq e$, we have

$$|u|_{\tau} = |u(e-e_1)|_{\tau} \leq |e-e_1|_{\tau} \leq |e-\rho^2|_{\tau} \leq \varepsilon$$

hence

$$|v-f|_{\tau} \leq |w-f|_{\tau} + |u|_{\tau} \leq 3\varepsilon .$$

The lemma is proved.

8.2 Let M be a von Neumann algebra and let e be a finite subfactor of M, with normalized trace τ. If $M = e \otimes (e' \cap M)$, we denote by $P_{e' \cap M}$ the faithful normal conditional expectation of M onto $e' \cap M$, which extends the map

$$x \otimes y \longrightarrow \tau(x)y , \qquad x \in e , \qquad y \in e' \cap M .$$

The following result is an immediate extension of Lemma 2.3.6 [4] of A. Connes, and is yielded by essentially the same proof.

LEMMA. *Let* M *be a factor and let* $e^1, e^2, \ldots, e^n, \ldots$ *be mutually commuting finite subfactors of* M, *such that* $M = e^n \otimes ((e^n)' \cap M)$ *for each* $n \geq 1$. *Suppose that for each* ϕ *in a total subset* Φ *of* M_* *we have*

$$\sum_{n \geq 1} \| \phi \circ P_{(e^n)' \cap M} - \phi \| < \infty \quad .$$

Then if e *denotes the weak closure of* $\bigcup_n e^n$ *in* M, e *is a finite subfactor of* M *and* $M = e \otimes (e' \cap M)$.

8.3 In all that follows, the group G that is dealt with will be assumed discrete and at most countable, and the factor M will be assumed to have a separable predual; ω will denote a free ultrafilter on $\mathbb{N}$.

LEMMA. *Let* G *be an amenable group and let* M *be a McDuff factor. Let* $\alpha: G \to \mathrm{Aut}\, M_\omega$ *be a semiliftable strongly free action. Then the fixed point algebra* $(M_\omega)^\alpha$ *is of the type* II_1.

Proof. Since M is a McDuff factor, M is of type II_1 by Theorem 5.2. Let I be a finite set, let $0 \in I$ and let $(e_{i,j})$, $i,j \in I$, be matrix units in M_ω. Then $e_{0,0} \sim \alpha_g(e_{0,0})$; so let $\bar{v}_g^0$ be a partial isometry in M_ω with $\bar{v}_g^{0*}\bar{v}_g^0 = \alpha_g(e_{0,0})$, $\bar{v}_g^0\bar{v}_g^{0*} = e_{0,0}$; for $g=1$ let $\bar{v}_1^0 = e_{0,0}$. Let us define the unitary

$$\bar{v}_g = \sum_i e_{i,0}\, \bar{v}_g^0 \alpha_g(e_{0,i}) \qquad g \in G$$

and let $((\bar{\alpha}_g),(\bar{u}_{g,h}))$ be the cocycle crossed action of G on M_ω obtained by perturbing the action (α_g) with $(\bar{v}_g)$. We infer for $i,j \in I$

$$\bar{\alpha}_g(e_{i,j}) = \bar{v}_g \alpha_g(e_{i,j}) \bar{v}_g^* = e_{i,0}\, \bar{v}_g^0 \alpha_g(e_{0,i} e_{i,j} e_{j,0}) \bar{v}_g^{0*} e_{0,j} = e_{i,j}$$

hence

$$\mathrm{Ad}\, \bar{u}_{g,h}(e_{i,j}) = \bar{\alpha}_g \bar{\alpha}_h \bar{\alpha}_{gh}^{-1}(e_{i,j}) = e_{i,j}$$

and $\bar{u}_{g,h} \in e' \cap M_\omega$, where e is the subfactor of M_ω generated by $(e_{i,j})$. We apply Proposition 7.4 to perturb $((\bar{\alpha}_g),(\bar{u}_{g,h}))$ with $(\tilde{v}_g) \subset e' \cap M_\omega$ to an action $(\tilde{\alpha}_g)$. Since $(v_g) = (\tilde{v}_g \bar{v}_g)$ perturbs the action (α_g) to the action $(\tilde{\alpha}_g)$, (v_g) is an (α_g) cocycle. Moreover,

$$\tilde{\alpha}_g(e_{i,j}) = \mathrm{Ad}\, \tilde{v}_g(\bar{\alpha}_g(e_{i,j})) = \mathrm{Ad}\, \tilde{v}_g(e_{i,j}) = e_{i,j} \quad .$$

We apply Proposition 7.2 to the (α_g) cocycle (v_g) and obtain a unitary $w \in M_\omega$ such that $v_g = w^*\alpha_g(w)$, $g \in G$.

Let us take $f_{i,j} = \mathrm{Ad}\, w(e_{i,j})$, $i,j \in I$. Then $(f_{i,j})$ are matrix units in M_ω and

$$\alpha_g(f_{i,j}) = \alpha_g(\mathrm{Ad}\, w(e_{i,j})) = \mathrm{Ad}(wv_g)(\alpha_g(e_{i,j}))$$
$$= \mathrm{Ad}\, w(e_{i,j}) = f_{i,j} \quad .$$

This ends the proof of the lemma.

8.4 By means of the lemma that follows we can lift constructions from M_ω to M.

LEMMA. *Let* $\alpha\colon G \to \mathrm{Aut}\, M$ *be a centrally free action of the amenable group* G *on the factor* M. *Let* $(V_g) \subset M^\omega$ *be a cocycle for* $(\alpha_g)^\omega$ *and let* $(E_{i,j})$, $i,j \in I$, $|I| < \infty$, *be matrix units in* M^ω *such that*

$$(\mathrm{Ad}\, V_g\alpha_g^\omega)(E_{i,j}) = E_{i,j} \qquad i,j \in I, \quad g \in G .$$

Then there exist representing sequences $(e_{i,j}^\nu)_\nu$ *for* $E_{i,j}$, *which for* $\nu \in N$ *are matrix units in* M, *and* $(v_g^\nu)_\nu$ *for* V_g, *which for each* ν *form an* (α_g)*-cocycle in* M, *such that*

$$(\mathrm{Ad}\, v_g^\nu\alpha_g)(e_{i,j}^\nu) = e_{i,j} \qquad i,j \in I, \quad g \in G, \quad \nu \in \mathbb{N} .$$

<u>Proof</u>. <u>Step A</u>. By Lemma 7.1 choose representing sequences $(e_{i,j}^\nu)_\nu$ for $E_{i,j}$ yielding for each ν matrix units in M. Let 0 be a distinguished element of I. For each $g \in G$, let $(\bar{v}_g^\nu)_\nu$ be a representing sequence for V_g consisting of unitaries in M, with $\bar{v}_1^\nu = 1$, $\nu \in \mathbb{N}$.

We have for all ν and g

$$(\mathrm{Ad}\, \bar{v}_g^\nu\alpha_g)(e_{0,0}^\nu) \sim e_{0,0}^\nu$$

and the sequences $((\mathrm{Ad}\, \bar{v}_g^\nu\alpha_g)(e_{0,0}^\nu))_\nu$ and $(e_{0,0}^\nu)_\nu$ both represent $(\mathrm{Ad}\, V_g\alpha_g^\omega)(E_{0,0}) = E_{0,0}$. By Lemma 7.1 there exists a sequence (w_g^ν) of partial isometries in M, representing $E_{0,0}$ and satisfying $w_g^{\nu *} w_g^\nu = (\mathrm{Ad}\, \bar{v}_g^\nu\alpha_g)(e_{0,0}^\nu)$, $w_g^\nu w_g^{\nu *} = e_{0,0}^\nu$; we take $w_1^\nu = e_{0,0}^\nu$. If we define unitaries $\bar{w}_g^\nu \in M$ by

$$\bar{w}_g^\nu = \sum_i e_{i,0}^\nu\, w_g^\nu (\mathrm{Ad}\, \bar{v}_g^\nu\alpha_g)(e_{0,i}^\nu)$$

then the sequence $(\bar{w}_g^\nu)_\nu$ represents

$$\sum_i E_{i,0}\, E_{0,0} (\mathrm{Ad}\, V_g\alpha_g^\omega)(E_{0,i}) = 1 \in M^\omega$$

and, moreover, as in the previous lemma, we infer

$$(\mathrm{Ad}(\bar{w}_g^\nu \bar{v}_g^\nu)\,\alpha_g)(e_{i,j}^\nu) = e_{i,j}^\nu \quad .$$

Hence $(\tilde{v}_g^\nu) = (\bar{w}_g^\nu \bar{v}_g^\nu)$ represents V_g and

$$(\mathrm{Ad}\ \tilde{v}_g\alpha_g)(e_{i,j}^\nu) = e_{i,j}^\nu \quad .$$

Step B. Let $\nu \in \mathbb{N}$ and let e^ν be the subfactor of M generated by $(e_{i,j}^\nu)_{i,j}$. Let $((\tilde{\alpha}_g^\nu),(\tilde{u}_{g,h}^\nu))$ be the cocycle crossed action obtained by perturbing the action (α_g) by $(\tilde{v}_g^\nu)$. Since $\tilde{\alpha}_g^\nu|e = \mathrm{id}$, we infer $\tilde{u}_{g,h}^\nu \in (e^\nu)'\cap M$; $g,h \in G$, $\nu \in \mathbb{N}$, and hence $((\tilde{\alpha}_g^\nu|(e^\nu)'\cap M),(\tilde{u}_{g,h}^\nu))$ is a cocycle crossed action of G on $(e^\nu)'\cap M$, which by **5.8** is centrally free. By Theorem 7.4, we can perturb $((\tilde{\alpha}_g^\nu),(\tilde{u}_{g,h}^\nu))$ with $(\tilde{w}_g^\nu) \subset (e^\nu)'\cap M$ to obtain an action (β_g^ν). Since the sequence $(\tilde{u}_{g,h}^\nu) = (\tilde{v}_g^\nu\alpha_g(\tilde{v}_h^\nu)\tilde{v}_{gh}^{\nu *})$ represents $V_g\alpha_g^\omega(V_h)V_{gh}^* = 1 \in M^\omega$, we have for each $g,h \in G$

$$\lim_{\nu\to\omega} \tilde{u}_{g,h}^\nu = 1 \qquad \text{*-strongly}$$

and by the estimates in the theorem, we may assume that $(\tilde{w}_g^\nu)$ also satisfies

$$\lim_{\nu\to\omega} \tilde{w}_g^\nu = 1 \qquad \text{*-strongly .}$$

We let $v_g^\nu = \tilde{w}_g^\nu\tilde{v}_g^\nu$. Since for each ν, (v_g^ν) perturbs the action (α_g) to an action (β_g), (v_g^ν) is an (α_g)-cocycle. For each $g \in G$, (v_g^ν) represents V_g, and for each $i,j \in I$

$$(\mathrm{Ad}\ v_g^\nu\alpha_g)(e_{i,j}^\nu) = \mathrm{Ad}\ \tilde{w}_g^\nu((\mathrm{Ad}\ \tilde{v}_g^\nu\alpha_g)(e_{i,j}^\nu)) = \mathrm{Ad}\ \tilde{w}_g^\nu(e_{i,j}^\nu) = e_{i,j}^\nu$$

and the lemma is proved.

8.5 The following result implies Theorem 1.2.

THEOREM. *Let* $\alpha\colon G \to \mathrm{Aut}\, M$ *be a centrally free action of the amenable group* G *on the McDuff factor* M. *Let* $\varepsilon > 0$, *let* Ψ *be a finite subset of* M_*^+ *and let* F *be a finite subset of* G. *Then there exists a cocycle* (v_g) *for* (α_g) *and a* II_1 *hyperfinite subfactor* $R \subset M$, *such that* $M = R \otimes (R'\cap M)$, $(\mathrm{Ad}\ v_g\alpha_g)|R = \mathrm{id}_R$ *and*

$$\|v_g - 1\|_\psi^\# < \varepsilon \qquad \psi \in \Psi,\quad g \in F\ ;$$

$$\|\psi \circ P_{R'\cap M} - \psi\| < \varepsilon \qquad \psi \in \Psi \quad .$$

In Theorem 1.2 we asserted that $(\mathrm{Ad}\ v_g\alpha_g|R'\cap M)$ is conjugate to

(α_g), and this can easily be obtained from the theorem above, since id_R is conjugate to $id_R \otimes id_R$.

<u>Proof</u>. We apply the previous lemma inductively to lift fixed point factors from M_ω to M.

Let $(F_n)_n$, $F_1 = F$, be an ascending sequence of finite subsets of G, with $\bigcup_n F_n = G$, and let $(\Psi_n)_n$, $\Psi_1 = \Psi$, be an ascending sequence of finite subsets of M_*^+, with $\bigcup_n \Psi_n$ total in M_*. We construct mutually commuting subfactors $\bar{e}^1, \bar{e}^2, \ldots, \bar{e}^n, \ldots$ of M, of type I_2, and cocycles $(\bar{v}_g^1)$ for $(\alpha_g^0) = (\alpha_g)$, $(\bar{v}_g^2)$ for $(\alpha_g^1) = (\mathrm{Ad}\, \bar{v}_g^1 \alpha_g^0), \ldots, (\bar{v}_g^{n+1})$ for $(\alpha_g^n) = (\mathrm{Ad}\, \bar{v}_g^n \alpha_g^{n-1}), \ldots$ such that if we let e^n be the subfactor of M generated by $\bar{e}^1 \cup \ldots \cup \bar{e}^n$, $e^0 = \mathbb{C}.1$, we have for each $n \geq 1$ $\alpha_g^n | e^{n-1} = id_{e^{n-1}}$ and $\bar{v}_g^n \;\; (e^{n-1})' \cap M$, and letting $v^n = \bar{v}_g^n \bar{v}_g^{n-1} \ldots \bar{v}_g$, $v_g^0 = 1$, we have

(1) $\quad \| v_g^n - v_g^{n-1} \|_\psi^\# \leq 2^{-n}\varepsilon \qquad g \in F_n, \quad \psi \in \Psi_n$;

(2) $\quad \| \psi \circ P_{(e^n)' \cap M} - \psi \| \leq 2^{-n}\varepsilon \qquad \psi \in \Psi_n$.

Let $n \geq 1$ and suppose, if $n > 1$, that $\bar{e}^1, \ldots, \bar{e}^{n-1}$ and $\bar{v}_g^1, \ldots, \bar{v}_g^{n-1}$ with the above properties have already been constructed. By **5.8** the factor $N = (e^{n-1})' \cap M$ is McDuff and $(\beta_g) = (\alpha_g^{n-1} | N)$ is a centrally free action of G on N. By Lemma 8.4 there are matrix units $(\tilde{E}_{i,j})$, $i,j \in \{0,1\}$ in $(N_\omega)^\beta$. By Lemma 8.3 (in which we take $(V_g) \equiv 1$), we may find representing sequences $(\tilde{e}_{i,j}^\nu)$ for $\tilde{E}_{i,j}$, consisting of matrix units in N, and for each ν an (α_g^{n-1}) cocycle $(\tilde{v}_g^\nu)$ in N such that

(3) $\quad (\mathrm{Ad}\, \tilde{v}_g^\nu \alpha_g^{n-1})(\tilde{e}_{i,j}^\nu) = \tilde{e}_{i,j}^\nu$

and $\lim_{\nu \to \omega} \tilde{v}_g^\nu = 1$ *-strongly.

For each $\psi \in N_*$, $\lim_{\nu \to \omega} \| [\tilde{e}_{i,j}^\nu \psi] \| = 0$. This also holds for each $\psi \in e_*^{n-1} \times N_* = M_*$. Let $\tilde{e}^\nu \subset M$ be the subfactor generated by $\tilde{e}_{i,j}^\nu$. We have

$$P_{(\tilde{e}^\nu)' \cap M}(x) = \tfrac{1}{2} \sum_{i,j} \tilde{e}_{i,j}^\nu \, x \, \tilde{e}_{j,i}^\nu \quad , \quad x \in M$$

hence for $\psi \in M_*$,

$$\lim_{\nu \to \omega} \psi \circ P_{(\tilde{e}^\nu)' \cap M} = \lim_{\nu \to \omega} \tfrac{1}{2} \sum_{i,j} \tilde{e}_{j,i}^\nu \, \psi \, \tilde{e}_{i,j}^\nu$$

$$= \lim_{\nu \to \omega} \tfrac{1}{2} \sum_{i,j} \tilde{e}_{j,i}^\nu \, \tilde{e}_{i,j}^\nu \psi = \psi \quad .$$

We may thus choose $\nu \in \mathbb{N}$ such that

$$\|\tilde{v}_g v_g^{n-1} - v_g^{n-1}\|_\psi^\# \leq 2^{-n}\varepsilon \qquad \psi \in \Psi_n\,, \quad g \in F_n$$

$$\|\psi \circ P_{(\tilde{e}^\nu)' \cap M} - \psi\| \leq 2^{-n}\varepsilon \qquad \psi \in \Psi_n\,.$$

If we take $\bar{v}_g^n = \tilde{v}_g^\nu$ and let $\bar{e}^n = \tilde{e}^\nu$, then the induction hypothesis is satisfied. From (1) we infer for $m \geq n \geq 0$

$$\|v_g^m - v_g^n\|_\psi^\# \leq 2^{-n}\varepsilon \qquad \psi \in \Psi_{n+1}\,, \quad g \in F_{n+1}\,.$$

Hence $v_g = \lim_{n\to\infty} v_g^m$ *-strongly exists and yields an (α_g) cocycle; moreover,

$$\|v_g - 1\|_\psi^\# \leq \varepsilon \qquad \psi \in \Psi = \Psi_1 \qquad g \in F = F_1\,.$$

We let R be the subfactor of M generated by $\bigcup_n \bar{e}^n$; by Lemma 8.2 R is a hyperfinite II_1 factor and splits M. We have

$$\|\psi \circ P_{R' \cap M} - \psi\| \leq \sum_{n\geq 1} \|(\psi \circ P_{(e^n)' \cap M} - \psi) \circ P_{(e^{n-1})' \cap M}\|$$

$$\leq \sum_{n\geq 1} 2^{-n}\varepsilon = \varepsilon\,.$$

For $m \geq n \geq 1$ we have

$$(\mathrm{Ad}\, v_g^n \alpha_g)|e^n = \alpha_g^m|e^n = \mathrm{id}_{e^n} \qquad g \in G\,,$$

thus at the limit when $m \to \infty$ and then $n \to \infty$ we infer $\mathrm{Ad}\, v_g \alpha_g | R = \mathrm{id}_R$, $g \in G$. The theorem is proved.

8.6 Let us recall Theorem 1.3 under a slightly different form.

THEOREM. *Let* $\alpha: G \to \mathrm{Aut}\, M$ *be a centrally free action of the amenable group* G *on the McDuff factor* M. *Let* $\varepsilon > 0$, *let* Ψ *be a finite subset of* M_*^+ *and let* F *be a finite subset of* G. *There exists a cocycle* (v_g) *for* α_g *and a* II_1 *hyperfinite subfactor* $R \subset M$, *such that* $M = R \otimes (R' \cap M)$, $(\mathrm{Ad}\, v_g \alpha_g)(R) = R$, $(\mathrm{Ad}\, v_g \alpha_g | R)$ *is conjugate to the model action* (4.5) *and*

$$\|v_g - 1\|_\psi^\# < \varepsilon \qquad \psi \in \Psi\,, \; g \in F$$

$$\|\psi \circ P_{R' \cap M} - \psi\| < \varepsilon \qquad \psi \in \Psi$$

From the above statement we may obtain the supplementary assertion of Theorem 1.3 that $(\mathrm{Ad}\, v_g \alpha_g | R' \cap M)$ is conjugate to (α_g), since the model action $(\alpha_g^{(0)})$ is conjugate to $(\alpha_g^{(0)} \otimes \alpha_g^{(0)})$ by construction.

The model action is an infinite tensor product of copies of the submodel action. The proof of the theorem will consist of an inductive application of the lemma that follows, which yields a copy of the submodel action.

LEMMA. *By the conditions of the theorem, there exists a cocycle* $(\bar{v}_g)$ *for* α_g *and a* II_1 *hyperfinite subfactor* $e \subset M$, *such that* $M = e \otimes (e' \cap M)$, $(\mathrm{Ad}\,\bar{v}_g\alpha_g)(e) = e$, $(\mathrm{Ad}\,\bar{v}_g\alpha_g|e)$ *is conjugate to the submodel action,* $(\mathrm{Ad}\,\bar{v}_g\alpha_g|e'\cap M)$ *is outer conjugate to* (α_g), *and*

$$\|v_g - 1\|_\psi^\# < \varepsilon \qquad \psi \in \Psi, \quad g \in F$$

$$\|\psi \circ P_{R'\cap M} - \psi\| < \varepsilon \qquad \psi \in \Psi$$

The proof of the lemma will occupy the next section. We first give the proof of the theorem.

We may suppose that Ψ consists of faithful states of M. Let $(\Psi_n)_{n\geq 1}$ be an ascending family of finite sets of normal states of M, with $\Psi_1 = \Psi$ and $\bigcup_n \Psi_n$ total in M_*, and let $(F_n)_{n\geq 1}$ be an ascending family of finite subsets of G, with $F_1 = F$ and $\bigcup_n F_n = G$.

We inductively construct mutually commuting hyperfinite II_1 subfactors $\bar{e}^1, \bar{e}^2, \dots$ of M, with $M = \bar{e}^n \otimes ((\bar{e}^n)' \cap M)$ for each n, and cocycles $(\bar{v}_g^1)$ for $(\alpha_g^0) = (\alpha_g)$, $(\bar{v}_g^2)$ for $(\alpha_g^1) = (\mathrm{Ad}\,\bar{v}_g^1\alpha_g^0), \dots, (\bar{v}_g^{n+1})$ for $(\alpha_g^n) = (\mathrm{Ad}\,\bar{v}_g^n\alpha_g^{n-1}), \dots$ such that if e^n is the subfactor of M generated by $\bar{e}^1 \cup \dots \cup \bar{e}^n$, $e^0 = \mathbb{C}.1$, and if $v_g^n = \bar{v}_g^n\bar{v}_g^{n-1} \dots \bar{v}_g$, $v_g^0 = 1$, then (1,n) $\alpha_g^n(\bar{e}^n) = \bar{e}^n$ and $(\alpha_g^n|\bar{e}^n)$ is conjugate to the submodel action

(2,n) $\quad(\alpha_g^n|(e^n)'\cap M)$ is outer conjugate to (α_g)

(3,n) $\quad\bar{v}_g^n \in (e^{n-1})' \cap M, \qquad g \in G$

(4,n) $\quad\|v_g^n - v_g^{n-1}\|_\psi^\# < 2^{-n}\varepsilon \qquad g \in F_k, \ \psi \in \Psi$

(5,n) $\quad\|\psi - \psi \circ P_{(e^n)'\cap M}\| < 2^{-n}\varepsilon \qquad \psi \in \Psi$

hold. Let $n \geq 1$ and suppose, if $n > 1$, that $\bar{e}^1, \dots, \bar{e}^{n-1}$ and $\bar{v}_g^1, \dots, \bar{v}_g^{n-1}$ satisfying (1,k)-(5,k) for $k = 1, \dots, n-1$ have been constructed. Let $N = (e^{n-1})' \cap M$.

Let us choose for each $\psi \in \Psi_n$ some $\chi_1, \dots, \chi_p \in e_*^{n-1}$ and $\phi_1, \dots, \phi_p \in N_*$ such that under the identification $M = e^{n-1} \otimes N$ we have

$$\|\psi - \sum_i \chi_i \otimes \phi_i\| \leq 2^{-n-2} \quad .$$

Let $\Phi \in N_*$ be the set of all those $\phi_1, \dots, \phi_p$ which appear in the above decomposition for some $\psi \in \Psi_n$, and let $\delta > 0$ be such that

$$\delta \sum_i \|\chi_i\| \, \|\phi_i\| \leq 2^{-n-2}$$

for all $\psi \in \Psi_n$.

The action $(\alpha_g^{n-1}|N)$ is by the induction hypothesis outer conjugate to (α_g). We apply to it the lemma in this section to obtain a II_1 hyperfinite subfactor $\bar{e}^n$ of N with $N = \bar{e}^n \otimes ((\bar{e}^n)' \cap N)$ and a cocycle $(\bar{v}_g^n)$ for (α_g^{n-1}) such that with $e^n = e^{n-1} \otimes \bar{e}^n \subset M$ and $(\alpha_g^n) = (\mathrm{Ad}\, \bar{v}_g^n \alpha_g^{n-1})$ we have $\alpha_g^n(\bar{e}^n) = \bar{e}^n$; $(\alpha_g^n\, e^n)$ is conjugate to the submodel action and $(\alpha_g^n|(\bar{e}^n)' \cap N)$ is outer conjugate to $(\alpha_g^{n-1}|N)$.

$$\|\bar{v}_g^n - 1\|_\psi^\# \leq 2^{-n-1}\varepsilon$$

$$\|\bar{v}_g^n - 1\|_{\psi_g}^\# \leq 2^{-n-1}\varepsilon$$

for $g \in F_n$, $\psi \in \Psi$, where $\psi_g = \mathrm{Ad}\, v_g^{n-1}$, and also

$$\|\phi - \psi \circ P_{(e^n)' \cap N}\| \leq \delta \|\phi\| \qquad \phi \in \Phi$$

Via the inequality 7.7(1), we infer

$$\|v_g^n - v_g^{n-1}\|_\psi^\# = \|(\bar{v}_g^n - 1)v_g^{n-1}\|_\psi^\# \leq 2^{\frac{1}{2}}(\|\bar{v}_g^n - 1\|_\psi^\# + \|\bar{v}_g^n - 1\|_{\psi_g}^\#)$$

$$\leq 2^{\frac{1}{2}} \cdot 2^{-n-1}\varepsilon \leq 2^{-n}\varepsilon \quad .$$

For $\psi \in \Psi_n$, with $\chi_1, \ldots, \chi_p \in e_*^{n-1}$ and $\phi_1, \ldots, \phi_p \in \Phi \subset N_*$ as chosen before, if let $\bar{\psi} = \sum_i \chi_i \otimes \phi_i \in M_*$, then

$$\|\psi - \bar{\psi}\| \leq 2^{-n-2}\varepsilon$$

$$\|\bar{\psi} - \psi \circ P_{(e^n)' \cap M}\| \leq \sum_i \|\chi_i\| \, \|\phi_i - \phi_i \circ P_{(e^n)' \cap N}\|$$

$$\leq \delta \sum_i \|\chi_i\| \, \|\phi_i\| \leq 2^{-n-2}\varepsilon$$

and hence

$$\|\psi - \psi \circ P_{(e^n)' \cap M}\| \leq \|\psi - \bar{\psi}\| + \|(\psi - \bar{\psi}) \circ P_{(e^n)' \cap M}\| + \|\bar{\psi} - \bar{\psi} \circ P_{(e^n)' \cap M}\|$$

$$\leq 3 \cdot 2^{-n-2}\varepsilon < 2^{-n}\varepsilon \quad .$$

The induction hypothesis is thus fulfilled.

From the conditions (4,n), for $m \geq n \geq 0$ we infer

$$\|v_g^m - v_g^n\|_\psi^\# \leq 2^{-n}\varepsilon \qquad g \in F_{n+1}, \quad \psi \in \Psi$$

therefore the limit $v_g = \lim_{\to \infty} v_g^n$ *-strongly exists and yields a unitary cocycle for α_g which satisfies

$$\|v_g - 1\|_\psi^\# < \varepsilon \qquad g \in F = F_1, \quad \psi \in \Psi \quad .$$

We let R be the subfactor of M generated by $\bigcup_n \bar{e}^n$. The conditions (5,n) show, in view of Lemma 8.2, that R splits M. For each $m \geq n \geq 1$ the action

$$\mathrm{Ad}\, v_g^m \alpha_g | \bar{e}^n = \mathrm{Ad}\, v_g^n \alpha_g | \bar{e}^n = \alpha_g^n | \bar{e}^n$$

is conjugate to the submodel action, hence $(\mathrm{Ad}\, v_g \alpha_g | \bar{e}^n)$ is conjugate to the submodel action, and thus $(\mathrm{Ad}\, v_g \alpha_g | R)$ is conjugate to the model action.

We have, for $\psi \in \Psi = \Psi_1$

$$\|\psi \circ P_{R' \cap M} - \psi\| \leq \sum_{n \geq 1} \|(\psi - \psi \circ P_{(\bar{e}^n)' \cap M}) \circ P_{(e^{n-1})' \cap M}\|$$
$$\leq \sum_{n \geq 1} 2^{-n}\varepsilon = \varepsilon \quad .$$

The theorem is proved.

8.7 The proof of Lemma 8.6, given in the sequel, is the crucial point of this chapter.

According to **4.4**, the submodel can be approximated by a system of almost equivariant matrix units, which form a finite dimensional submodel product with a hyperfinite II_1 factor almost fixed by the action. In Step A below, we construct an almost equivariant system of matrix units in M. In Steps B and C, we perturb the action in order to make the almost equivariant s.m.u. become equivariant. In Step D we lift the whole construction from M_ω to M, and in Step E we construct the remaining almost invariant part of the submodel.

Throughout the proof we shall use the notations connected to the Paving Structure for G (3.4) on which the construction of the model action (4.4(5)) was based. Recall that $\varepsilon_n > 0$, $G_n \subset\subset G$, (K_i^n), $i \in I_n$ are the ε_n-paving, (ε_n, G_n) invariant sets on the n-th level of the Paving Structure, and $\ell_g^n\colon \bigcup_i K_i^n \to \bigcup_i K_i^n$ are bijections approximating the left g-translations. The assumptions on $(\varepsilon_n)_n$ done in 3.5 and based upon the fact that ε_{n+1} could be chosen very small with respect to $\varepsilon_0, \ldots, \varepsilon_n$, are used without further mention. Also recall that the set S_i^n is the multiplicity with which K_i^n enters in the construction of the submodel (see 4.4) and $\bar{S}^n = \bigcup_i K_i^n \times S_i^n$.

Let us choose $n \geq 4$ such that $30\varepsilon_{n-4}^{1/2} < \varepsilon$ and $G_{n-4} \supseteq F$.

Step A. The Rohlin Theorem provides an almost equivariant partition of unity in M ; from this together with a fixed point s.m.u. in M_ω we obtain, by diagonal summation, an almost equivariant s.m.u. in M_ω.

Lemma 5.6 shows that the action $(\alpha_g)_\omega$ induced by (α_g) on M_ω is strongly free. For simplicity of notation, we shall denote $(\alpha_g)_\omega$ by (α_g) as well. Since M is McDuff, by Lemma 8.3 the fixed point algebra $(M_\omega)^\alpha$ is of type II. We choose a s.m.u. (F_{s_1,s_2}), s_1,s_2 $\bar{S}^n$ in $(M_\omega)^\alpha$.

We apply the Rohlin Theorem 6.1 and get a partition of unity $(\bar{F}_{i,k})$, $i \in I_{n-1}$, $k \in K_i^{n-1}$ in M_ω such that

$$\sum_i \sum_{k,\ell} |\alpha_{k\ell^{-1}}(\bar{F}_{i,\ell}) - \bar{F}_{i,k}|_\tau < 5\varepsilon_{n-1}^{1/2}$$

$$[\alpha_g(\bar{F}_{i,k}), \bar{F}_{j,m}] = 0$$

$$[\bar{F}_{i,k}, F_{s_1,s_2}] = 0$$

for $i,j \in I_{n-1}$, $k,\ell \in K_i^{n-1}$, $m \in K_j^{n-1}$, $s_1,s_2 \in \bar{S}^n$.

We define a s.m.u. $(\bar{E}_{s_1,s_2})$, $s_1,s_2 \in \bar{S}^n$ in M_ω by

$$\bar{E}_{(k_1,s_1),(k_2,s_2)} = \sum_{i,h} F_{(\ell_1,s_1),(\ell_2,s_2)} \bar{F}_{i,h}$$

for $(k_1,s_1),(k_2,s_2) \in \bar{S}^n = \bigcup_j K_j^n \times S_j^n$, $i \in I_{n-1}$, $h \in K_i^{n-1}$ and $\ell_1 = \ell_{h^{-1}}^{n-1}(k_1)$, $\ell_2 = \ell_{h^{-1}}^{n-1}(k_2)$.

Since F_{s_1,s_2} and $\bar{F}_{i,k}$ commute and ℓ_g^n are bijections, it is easy to see that

$$(E_{(k_1,s_1),(k_2,s_2)} \bar{F}_{i,h})$$

form a s.m.u. under $\bar{F}_{i,h}$, for each fixed i,h; hence $(\bar{E}_{s_1,s_2})$ are a s.m.u.

Let us take

$$\tilde{S}^n = \{(k,s) \in \bar{S}^n \mid i \in I_n,\ k \in K_i^n \cap \bigcap_{g \in G_n} g^{-1}K_i^n,\ s \in S_i^n\} .$$

Since K_i^n is (ε_n, G_n) invariant, we have

$$(1) \quad |\tilde{S}^n| \geq (1 - \varepsilon_n)|\bar{S}^n| .$$

Let us keep $g \in G_{n-1}$; $(k_1,s_1),(k_2,s_2) \in \tilde{S}^n$ fixed. We have

$$\alpha_g(\bar{E}_{(k_1,s_1),(k_2,s_2)}) = \sum_{i,h} F_{(h^{-1}k_1,s_1),(h^{-1}k_2,s_2)} \alpha_g(\bar{F}_{i,h})$$

$$= \Sigma_1 + \Sigma_2$$

where $i \in I_{n-1}$, $h \in K_i^{n-1}$; in Σ_1 we sum for (i,h) with $h \in K_i^{n-1} \cap g^{-1}K_i^{n-1}$ and in Σ_2 for the rest of (i,h). On the other hand, we infer

$$\bar{E}_{(gk_1,s_1),(gk_2,s_2)} = \sum_{i,k} F_{(k^{-1}gk_1,s_1),(k^{-1}gk_2,s_2)} \bar{F}_{i,k}$$

$$= \Sigma_1' + \Sigma_2'$$

where $i \in I_{n-1}$, $k \in K_i^{n-1}$; in Σ_1' we sum for (i,k) with $k \in gK_i^{n-1} \cap K_i^{n-1}$ and in Σ_2' for the other (i,k). Since K_i^{n-1} is $(\varepsilon_{n-1},G_{n-1})$ invariant, we have for each $i \in I_{n-1}$,

$$|K_i^{n-1} \cap g^{-1}K_i^{n-1}| \geq (1-\varepsilon_{n-1})|K_i^{n-1}|$$

and so, by the estimates 6.1(5) for the Rohlin Theorem, we infer

$$|\Sigma_2|_\tau \leq |\bar{S}^n|^{-1} \sum_{i,h} |\bar{F}_{i,h}|_\tau$$

$$\leq |\bar{S}^n|^{-1}(5\varepsilon_{n-1}+\varepsilon_{n-1}) \leq 6\varepsilon_{n-1}|\bar{S}^n|^{-1}$$

for $h \in K^{n-1} \cap g^{-1}K^{n-1}$, and similarly

$$|\Sigma_2'|_\tau \leq 6\varepsilon_{n-1}|\bar{S}^n|^{-1} .$$

If we let $k = gh$ in Σ_1 we obtain

$$\Sigma_1 - \Sigma_1' = \sum_{i,h} F_{(h^{-1}k_1,s_1),(h^{-1}k_2,s_2)}(\alpha_g(\bar{F}_{i,h}) - \bar{F}_{i,gh})$$

where $i \in I_{n-1}$ and $h \in K_i^{n-1} \cap g^{-1}K_i^{n-1}$. The estimates 6.1(6) for the Rohlin partition $\bar{F}_{i,h}$ yield

$$|\Sigma_1 - \Sigma_1'| \leq |\bar{S}^n|^{-1} \sum_{i,h} |\alpha_g(\bar{F}_{i,h}) - \bar{F}_{i,gh}|_\tau$$

$$\leq |\bar{S}^n|^{-1}(5\varepsilon_{n-1}+5\varepsilon_{n-1}) \leq 10\varepsilon_{n-1}|\bar{S}^n|^{-1} .$$

Summing up, for $g \in G_{n-1}$ and $(k_1,s_1),(k_2,s_2) \in \tilde{S}^n \subseteq \bar{S}^n$, we have

(2) $$|\alpha_g(\bar{E}_{(k_1,s_1),(k_2,s_2)}) - \bar{E}_{(gk_1,s_1),(gk_2,s_2)}|_\tau$$
$$|\Sigma_1 - \Sigma_1'|_\tau + |\Sigma_2|_\tau + |\Sigma_2'|_\tau \leq 22\varepsilon_{n-1}|\bar{S}^n|^{-1} .$$

Step B. We perturb the action (α_g) with $(\bar{W}_g)$ to make it coincide on $(\bar{E}_{s_1,s_2})$ with a copy $(\mathrm{Ad}\,\bar{U}_g)$ of the n-th finite dimensional submodel.

For $g \in G$, let $\bar{U}_g$ be the unitary associated to the s.m.u. $(\bar{E}_{s_1,s_2})$, $s_1,s_2 \in \bar{S}^n$ in the same way as in the n-th finite dimensional submodel 4.4, i.e.

$$\bar{U}_g = \sum_{k,s} \bar{E}_{(k_g,s),(k,s)}$$

where $(k,s) \in \bar{S}^n$ and $k_g = \ell^n_g(k)$. Let (k_0,s_0) be some distinguished element of $\tilde{S}^n$, and let us choose for every $g \in G$, a partial isometry $\bar{W}^0_g$ such that

$$\bar{W}^{0*}_g \bar{W}^0_g = \alpha_g(\bar{E}_{(k_0,s_0),(k_0,s_0)})$$

$$W^0_g W^{0*}_g = \mathrm{Ad}\,\bar{U}_g(\bar{E}_{(k_0,s_0),(k_0,s_0)}) = \bar{E}_{(gk_0,s_0),(gk_0,s_0)}$$

$$\bar{W}^0_1 = \bar{E}_{(k_0,s_0),(k_0,s_0)} \quad .$$

According to Lemma 8.1.2, from (2) above we may assume that for $g \in G_{n-1}$ we have

$$(3) \quad |\bar{W}^0_g - \bar{E}_{(gk_0,s_0),(gk_0,s_0)}|_\tau \leq 66\varepsilon^{\frac{1}{2}}_{n-1}|\bar{S}^n|^{-1} \quad .$$

Let us define the unitary

$$\bar{W}_g = \sum_{i,s} \mathrm{Ad}\,\bar{U}_g(\bar{E}_{(k,s),(k_0,s_0)})\bar{W}^0_g\alpha_g(\bar{E}_{(k_0,s_0),(k,s)})$$

where $(k,s) \in \bar{S}^n$. From the definition we infer

$$(4) \quad (\mathrm{Ad}\,\bar{W}_g\alpha_g)(\bar{E}_{s_1,s_2}) = \mathrm{Ad}\,\bar{U}_g(\bar{E}_{s_1,s_2})$$

for any $g \in G$, $s_1,s_2 \in \bar{S}^n$. We estimate for $g \in G_{n-1}$

$$\bar{W}_g - 1 = \sum_{k,s}\Big(\mathrm{Ad}\,\bar{U}_g(\bar{E}_{(k,s),(k_0,s_0)})\bar{W}^0_g\alpha_g(\bar{E}_{(k_0,s_0),(k,s)}) - \mathrm{Ad}\,\bar{U}_g(E_{(k,s),(k,s)})\Big)$$

$$= \Sigma_1 + \Sigma_2$$

where $(k,s) \in \bar{S}^n$; in Σ_1 we sum for $(k,s) \in \tilde{S}^n$ and in Σ_2 for $(k,s) \in \bar{S}^n \setminus \tilde{S}^n$. In view of the estimate (1) on $\tilde{S}^n$, we have

$$|\Sigma_2| \leq 2|\bar{S}^n \setminus \tilde{S}^n|\,|\bar{S}^n|^{-1} \leq 2\varepsilon_{n-1} \quad .$$

For $(k,s) \in \tilde{S}^n$, the norm of the corresponding term in Σ_1 is

$$\Big|\bar{E}_{(gk,s),(gk_0,s_0)}\bar{W}^0_g\alpha_g(\bar{E}_{(k_0,s_0),(k,s)}$$
$$- \bar{E}_{(gk,s),(gk_0,s_0)}\bar{E}_{(gk_0,s_0),(gk_0,s_0)}\bar{E}_{(gk_0,s_0),(gk,s)}\Big|_\tau \leq$$

$$\leq |\bar{W}_g^0 - \bar{E}_{(gk_0,s_0),(gk_0,s_0)}|_\tau + |\alpha_g(\bar{E}_{(k_0,s_0),(k,s)}) - \bar{E}_{(gk_0,s_0),(gk,s)}|_\tau$$

$$\leq 66\varepsilon_{n-1}^{\frac{1}{2}}|\bar{S}^n|^{-1} + 22\varepsilon_{n-1}^{\frac{1}{2}}|\bar{S}^n|^{-1} = 88\varepsilon_{n-1}^{\frac{1}{2}}|\bar{S}^n|^{-1}$$

where for the last inequality we have used (2) and (3). Hence $|\Sigma_1|_\tau \leq 88\varepsilon_{n-1}$ and thus for any $g \in G_{n-1}$,

$$(5) \quad |\bar{W}_g - 1|_\tau \leq |\Sigma_1|_\tau + |\Sigma_2|_\tau \leq 88\varepsilon_{n-1}^{\frac{1}{2}} + 2\varepsilon_{n-1} \leq 90\varepsilon_{n-1}^{\frac{1}{2}} .$$

<u>Step C</u>. We use stability results to further perturb $(\mathrm{Ad}\, \bar{W}_g\alpha_g)$ with $(\tilde{W}_g)$, such that it continues to coincide on $(\bar{E}_{s_1,s_2})$ with $\mathrm{Ad}\, \bar{U}_g$, but, moreover, $(\bar{U}_g^*\tilde{W}_g\bar{W}_g)$ is an (α_g)-cocycle.

Let $\bar{E} \subset M_\omega$ be the subfactor generated by $(\bar{E}_{s_1,s_2})$. Let us consider the cocycle crossed action $((\bar{\alpha}_g),(\bar{Z}_{g,h}))$ of G on M_ω, obtained by perturbing the action (α_g) with $(\bar{U}_g^*\bar{W}_g)$. We have from (4)

$$\mathrm{Ad}(\bar{U}_g^*\bar{W}_g)\alpha_g|\bar{E} = \mathrm{id}_{\bar{E}}$$

and since $\bar{U}_g \in \bar{E}$, we infer

$$\begin{aligned}
\bar{Z}_{g,h} &= \bar{U}_g^*\bar{W}_g\alpha_g(\bar{U}_h^*\bar{W}_h)\bar{W}_{gh}^*\bar{U}_{gh} \\
&= \mathrm{Ad}(\bar{U}_g^*\bar{W}_g)(\alpha_g(\bar{U}_h^*))\ \bar{U}_g^*\bar{W}_g\alpha_g(\bar{W}_h)\bar{W}_{gh}^*\bar{U}_{gh} \\
&= \bar{U}_h^*\bar{U}_g^*\bar{W}_g\alpha_g(\bar{W}_h)\bar{W}_{gh}^*\bar{U}_{gh} = \bar{U}_h^*\bar{U}_g^*\bar{U}_{gh}\ \mathrm{Ad}\ \bar{U}_{gh}^*(Z_{g,h})
\end{aligned}$$

where $Z_{g,h} = \bar{W}_g\alpha_g(\bar{W}_h)\bar{W}_{gh}^*$. For $g,h,gh \in G_{n-1}$ we have from (5)

$$|Z_{g,h} - 1|_\tau \leq |\bar{W}_g - 1|_\tau + |\bar{W}_h - 1|_\tau + |\bar{W}_{gh} - 1|_\tau \leq 270\varepsilon_{n-1}^{\frac{1}{2}}$$

and since $((\bar{E}_{s_1,s_2}),(\bar{U}_g))$ is isomorphic to the n-th finite dimensional submodel, the inequality 4.4(3) yields for $g,h,gh \in G_{n-1}$

$$|\bar{U}_{gh} - \bar{U}_g\bar{U}_h|_\tau \leq 2\varepsilon_n .$$

Hence for $g,h,gh \in G_{n-1}$ we obtain

$$\begin{aligned}
|\bar{Z}_{g,h} - 1|_\tau &\leq |\bar{U}_g^*\bar{U}_h^*\bar{U}_{gh} - 1|_\tau + |\mathrm{Ad}\ \bar{U}_{gh}^*(Z_{g,h} - 1)|_\tau \\
&\leq 2\varepsilon_n + 270\varepsilon_{n-1}^{\frac{1}{2}} \leq \varepsilon_{n-4} .
\end{aligned}$$

Since $\bar{\alpha}_g|\bar{E} = \mathrm{id}_{\bar{E}}$, we have

$$\mathrm{Ad}\ \bar{Z}_{g,h}|\bar{E} = (\bar{\alpha}_g\bar{\alpha}_h\bar{\alpha}_{gh}^{-1})|\bar{E} = \mathrm{id}_{\bar{E}}$$

hence $(\bar{Z}_{g,h}) \subset \bar{E}' \cap M_\omega$. We apply Proposition 7.4 to obtain

$(\tilde{W}_g) \subset E' \cap M_\omega$ perturbing the cocycle crossed action $((\bar{\alpha}_g),(\bar{z}_{g,h}))$ to an action (α_g) such that

(6) $|\tilde{W}_g - 1|_\tau \leq 18\varepsilon_{n-4}$ for $g \in G_{n-4}$.

Since $\bar{U}_g \in \bar{E}$ commutes with $\tilde{W}_g$, we define

$$W_g = \tilde{W}_g \bar{U}_g^* \bar{W}_g = \bar{U}_g^* \tilde{W}_g \bar{W}_g$$

and infer

$$(\mathrm{Ad}\, W_g \alpha_g)|\bar{E} = (\mathrm{Ad}\, \tilde{W}_g \bar{\alpha}_g)|\bar{E} = \mathrm{Ad}\, \tilde{W}_g|\bar{E} = \mathrm{id}_{\bar{E}} .$$

Since (W_g) perturbs the action (α_g) to the action $(\tilde{\alpha}_g)$, it is an (α_g)-cocycle. We have from (5) and (6) for $g \in G_{n-4}$,

$$|\bar{U}_g W_g - 1|_\tau^\# = |\tilde{W}_g \bar{W}_g - 1|_\tau \leq |\tilde{W}_g - 1|_\tau + |\bar{W}_g - 1|_\tau \leq 18\varepsilon_{n-4} + 90\varepsilon_{n-1}^{\frac{1}{2}} \leq 19\varepsilon_{n-4}$$

and so

$$\|\bar{U}_g W_g - 1\|_\tau^\# = \|\bar{U}_g W_g - 1\|_\tau \leq (|\bar{U}_g W_g - 1|_\tau \|\bar{U}_g W_g - 1\|)^{\frac{1}{2}} \leq (38\varepsilon_{n-4})^{\frac{1}{2}} < 7\varepsilon_{n-4}^{\frac{1}{2}} .$$

Step D. We lift the construction done before from M_ω to M. Let us apply Lemma 8.4 to the action $(\mathrm{Ad}\, W_g \alpha_g)$, which keeps $(\bar{E}_{s_1,s_2})$ fixed. Let $(\bar{e}_{s_1,s_2})$ be s.m.u. in M representing $(\bar{E}_{s_1,s_2})$ and let $(w_g^\nu)_\nu$ be (α_g) cocycles in M representing (W_g) such that for each $\nu \in N$

$$(\mathrm{Ad}\, w_g^\nu \alpha_g)|\bar{e} = \mathrm{id}_{\bar{e}^\nu}$$

where $\bar{e}^\nu$ is the subfactor of M generated by $(\bar{e}^\nu_{s_1,s_2})$. We define

$$\bar{u}_g^\nu = \sum_{k,s} \bar{e}^\nu_{(k_g,s),(k,s)}$$

where $i \in I_n$, $(k,s) \in K^n \times S^n \subseteq \bar{S}^n$ and $k_g = \ell_g^n(k)$, which, compared with the definition of $\bar{U}_g$, shows that $(\bar{u}_g^\nu)$ represents $\bar{U}_g$.

For any $\phi \in M_*$ we have

$$\lim_{\nu \to \omega} \|\phi \circ P_{(\bar{e}^\nu)' \cap M} - \phi\| = \lim_{\nu \to \omega} \Big\| |\bar{S}^n|^{-1} \sum_{s_1,s_2} (\bar{e}^\nu_{s_1,s_2} \phi \bar{e}^\nu_{s_1,s_2} - \bar{e}^\nu_{s_1,s_2} \bar{e}^\nu_{s_2,s_1} \phi) \Big\|$$

$$\lim_{\nu \to \omega} |\bar{S}^n|^{-1} \sum_{s_1,s_2} \|[\phi, \bar{e}^\nu_{s_1,s_2}]\| = 0 .$$

By choosing $\nu \in \mathbb{N}$ in a suitable way we may thus obtain a s.m.u. $(\bar{e}_{s_1,s_2})$, $s_1,s_2 \in \bar{S}^n$ in M, generating a subfactor $\bar{e}$ of M, a unitary cocycle (w_g) for (α_g) and unitaries $(\bar{u}_g)$ in $\bar{e}$ such that $(\bar{e},\bar{u}_g)$ is a copy of the n-th finite dimensional submodel and

$$\mathrm{Ad}\ w_g\alpha_g|\bar{e} = \mathrm{id}_{\bar{e}}$$

$$\|\bar{u}_g w_g - 1\|_\psi^\# \leqslant 7\varepsilon_{n-4} \qquad g \in G_{n-4},\quad \psi \in \Psi$$

$$\|\phi \circ P_{\bar{e}'\cap M} - \phi\| \leqslant 1/8 \qquad \phi \in \Phi \cup \Psi$$

Step E. We complete the finite dimensional submodel $\bar{e}$ with a subfactor f of M which is almost fixed by (α_g), to obtain a copy of the submodel.

Let $N = \bar{e}'\cap M$. The restriction of the action $(\mathrm{Ad}\ w_g\alpha_g)$ to N by **5.8** is centrally free, and thus we may obtain, by Theorem 1.2, a hyperfinite II_1 subfactor f of N, with $N = f \otimes (f'\cap N)$ and a cocycle $(z_g) \subset N$ for $(\mathrm{Ad}\ w_g\alpha_g)$ such that

$$\mathrm{Ad}(z_g w_g)\alpha_g|f = \mathrm{id}_f$$

$$(\mathrm{Ad}(z_g w_g)\alpha_g \mid f'\cap N) \quad \text{is conjugate to}$$
$$\mathrm{id}_{\bar{e}} \otimes (\mathrm{Ad}\ w_g\alpha_g|N) = (\mathrm{Ad}\ w_g\alpha_g)$$

$$\|z_g - 1\|_\psi^\# \leqslant 1/8\,\varepsilon \qquad g \in G_{n-4},\quad \psi \in \Psi|N$$

$$\|\phi \circ P_{f'\cap N} - \phi\| \leqslant 1/8\,\varepsilon \qquad \phi \in (\Phi \cup \Psi)|N .$$

The subfactor e of M generated by $\bar{e} \cup f$ is isomorphic to the factor on which the submodel action acts, and $M = e \otimes (e'\cap M)$. If we choose an isomorphism between e and the submodel coinciding on $\bar{e}$ with the one chosen in the previous step, we get a unitary representation (u_g) of G into e, copy of the model representation, such that from 4.4.(1),

$$|u_g - \bar{u}_g|_{\bar{\psi}} \leqslant 8\varepsilon_n \qquad g \in G_n$$

for $\psi \in \Psi$, $\bar{\psi} = \psi \circ P_{e'\cap M}$, since $\bar{\psi}|e$ is the normalized trace. This yields for $g \in G_n$,

$$|u_g\bar{u}_g - 1|_{\bar{\psi}} \leqslant 8\varepsilon_n$$

$$\|u_g\bar{u}_g^* - 1\|_{\bar{\psi}}^\# \leqslant (|u_g\bar{u}_g^* - 1|_{\bar{\psi}}\,\|u_g\bar{u}_g^* - 1\|)^{1/2} \leqslant 4\varepsilon_n^{1/2} .$$

For any $x \in M$ and any normal states $\psi, \bar{\psi} \in M_*$

$$\|x\|_\psi^{\#2} \leqslant \|x\|_{\bar{\psi}}^{\#2} + \|\psi - \bar{\psi}\|\,\|x\|^2$$

hence

Since id_R *is conjugate to* $\mathrm{id}_R \otimes \mathrm{id}_R$, *one can actually assume that moreover*

$$(\mathrm{Ad}\, v_g \alpha_g | R' \cap M) \quad \textit{is conjugate to} \quad (\alpha_g) \quad .$$

Towards this result one first proves the following analogue of Lemma 8.4.

LEMMA. *Let* $\alpha: G \to \mathrm{Aut}\, M$ *be a centrally free crossed action of the amenable group* G *on the factor* M. *Let* $(v_g) \subset M^\omega$ *be unitaries, with* $v_1 = 1$, *and let* $(E_{i,j})$, $i,j \in I$, $|I| < \infty$, *be a s.m.u. in* M^ω *such that*

$$(\mathrm{Ad}\, v_g \alpha_g^\omega)(E_{i,j}) = E_{i,j} \qquad i,j \in I\,, \quad g \in G \quad .$$

Then there exist representing sequences of s.m.u. $(e_{i,j}^\nu)_\nu$ *for* $(E_{i,j})$, *and representing sequences of unitaries* $(v_g^\nu)_\nu$ *for* V_g, $v_1^\nu = 1$, *such that*

$$(\mathrm{Ad}\, v_g^\nu \alpha_g)(e_{i,j}^\nu) = e_{i,j}^\nu \qquad i,j \in I\,, \quad g \in G\,, \quad \nu \in N\,.$$

The proof of this lemma consists merely of Step A of the proof of Lemma 8.4 (actually the group property of G is not needed). The proof of the theorem is obtained from the one of Theorem 8.5 by using the above lemma instead of Lemma 8.4. Since a crossed action induces an action on the centralizing algebra M_ω (because inner automorphisms are centrally trivial), Lemma 8.3 can still be used.

8.9 Theorem 1.6 is implied by the following result.

THEOREM. *Let* $\alpha: G \to \mathrm{Aut}\, M$ *be a centrally free crossed action of the amenable group* G *on the McDuff factor* M. *Let* $\varepsilon > 0$, *let* Ψ *be a finite subset of* M_*^+ *and let* F *be a finite subset of* G. *Then there exists a family* (v_g), $g \in G$, *of unitaries in* M, $v_1 = 1$, *and a* II_1 *hyperfinite subfactor* $R \subset M$, *such that* $M = R \otimes (R' \cap M)$, $(\mathrm{Ad}\, v_g \alpha_g)(R) = R$, $(\mathrm{Ad}\, v_g \alpha_g | R)$ *is conjugate to the model action and*

$$\| v_g - 1 \|_\psi^\# < \varepsilon \qquad \psi \in \Psi, \quad g \in F$$

$$\| \psi \circ P_{R' \cap M} - \psi \| < \varepsilon \qquad \psi \in \Psi\,.$$

Once again, the supplementary assertion that $(\mathrm{Ad}\, v_g \alpha_g | R' \cap M)$ be conjugate to (α_g) may be obtained since the model action $(\alpha_g^{(0)})$ is conjugate to $(\alpha_g^{(0)} \otimes \alpha_g^{(0)})$.

The proof is similar to the one of Theorem 8.6. Since (α_g) induces an action on M_ω, Steps A, B and C remain unchanged. In Step D the

(7) $\|x\|^{\#}_{\psi} \leqslant \|x\|^{\#}_{\bar\psi} + \|\psi - \bar\psi\|^{\frac12}\|x\|$.

Letting $x = u_g\bar u_g^* - 1$, $\psi \in \Psi$ and $\bar\psi = \psi \circ P_{e' \cap M}$, we have

$$\|\psi - \bar\psi\| \leqslant \|\psi - \psi \circ P_{\bar e' \cap M}\| + \|(\psi - \psi \circ P_{f' \cap M}) \circ P_{\bar e' \cap M}\|$$

$$\leqslant 1/8\,\varepsilon + 1/8\,\varepsilon = 1/4\,\varepsilon$$

we infer

$$\|u_g\bar u_g^* - 1\|_\psi \leqslant \|u_g u_g^* - 1\|_{\bar\psi} + \|\psi - \bar\psi\|^{\frac12}\|u_g\bar u_g^* - 1\| \leqslant 4\varepsilon_n^{\frac12} + 1/4\,\varepsilon .$$

Since (u_g) is a representation kept fixed by the action $(\mathrm{Ad}(z_g w_g)\alpha_g)$, (u_g) is an $(\mathrm{Ad}(z_g w_g)\alpha_g)$-cocycle. Therefore, if we let $\bar v_g = u_g z_g w_g$, then $(\bar v_g)$ is an (α_g) cocycle. The action $(\mathrm{Ad}\ \bar v_g\alpha_g)$ leaves e globally invariant and coincides on e with $(\mathrm{Ad}\ u_g)$, the copy of the submodel action. Since $z_g \in \bar e' \cap M$ and $\bar u_g \in \bar e$, we infer

$$\bar v_g = u_g z_g \bar u_g^* \bar u_g w_g = u_g \bar u_g^* z_g \bar u_g w_g$$

and hence, via the inequality 7.1(10) we obtain

$$\|\bar v_g - 1\|^{\#}_{\psi} \leqslant 2(\|u_g\bar u_g^* - 1\|^{\#}_{\psi} + \|z_g - 1\|^{\#}_{\psi} + \|\bar u_g w_g - 1\|^{\#}_{\psi})$$

$$2((4\varepsilon_n^{\frac12} + \tfrac14\varepsilon) + 1/8\,\varepsilon + 7\varepsilon_{n-4}^{\frac12}) \leqslant 15\varepsilon_{n-4}^{\frac12} + \frac{\varepsilon}{2} \leqslant \varepsilon$$

for any $g \in F \subseteq G_{n-4}$.

The proof of the lemma is finished.

8.8 It will be convenient for the proofs, instead of dealing with G-kernels, which are homomorphisms $G \to \mathrm{Out}\, M = \mathrm{Aut}\, M/\mathrm{Int}\, M$, to work with their sections, which we called crossed actions, and which are maps $\alpha: G \to \mathrm{Aut}\, M$, $\alpha_1 = \mathrm{id}$, with $\alpha_g\alpha_h\alpha_{gh}^{-1} \in \mathrm{Int}\, M$ for $g,h \in G$. The following theorem implies Theorem 1.5.

THEOREM. *Let* $\alpha: G \to \mathrm{Aut}\, M$ *be a centrally free crossed action of the amenable group* G *on the McDuff factor* M. *Let* $\varepsilon > 0$, *let* Ψ *be a finite subset of* M_*^+ *and let* F *be a finite subset of* G. *There exist unitaries* $v_g \in G$, $g \in G$, *with* $v_1 = 1$, *and a* II_1 *hyperfinite subfactor* $R \subset M$ *such that* $M = R \otimes (R' \cap M)$, $(\mathrm{Ad}\ v_g\alpha_g)|R = \mathrm{id}_R$ *and*

$$\|v_g - 1\|^{\#}_{\psi} < \varepsilon \qquad \psi \in \Psi,\ g \in F$$

$$\|\psi \circ P_{R' \cap M} - \psi\| < \varepsilon \qquad \psi \in \Psi .$$

the lemma in the preceding section is used instead of Lemma 8.4 and in Step E, Theorem 8.5 is replaced by Theorem 8.8.

Chapter 9: MODEL ACTION ISOMORPHISM

In this chapter we give the proof of the main result of this paper, Theorem 1.4, which characterizes the centrally free actions which are approximately inner.

9.1 In this section we implement V. Jones' idea of dealing with approximate inner actions of a group G by means of a $G \times G$ action.

Throughout this chapter we again assume that the group G is discrete, at most countable, and that the factor M has a separable predual; we let ω be a free ultrafilter on $\mathbb{N}$.

LEMMA. *Let* $\alpha: G \to \operatorname{Aut} M$ *be a centrally free approximately inner action of the amenable group* G *on the factor* M. *For each* $g \in G$, *let* $(v_g^\nu)_\nu$ *be a sequence of unitaries in* M *such that* $\alpha_g = \lim_{\nu\to\omega} \operatorname{Ad} v_g^\nu$, *and let* $V_g = (v_g^\nu)_\nu \in M^\omega$; $V_1 = 1$.

Let us take, for $g,h,k,\ell \in G$,

$$\theta_{(g,h)} = \operatorname{Ad} V_{gh^{-1}}\alpha_h^\omega \in \operatorname{Aut} M^\omega$$

$$U_{(g,h),(k,\ell)} = V_{gh^{-1}}\alpha_h(V_{k\ell^{-1}})V^*_{gk\ell^{-1}h^{-1}} \in M^\omega \quad .$$

Then $(\theta|M_\omega, U)$ *is a cocycle crossed action of* $G \times G$ *on* M_ω, *which is semiliftable and strongly free* (see **5.2, 5.6**).

<u>Proof</u>. The fact that each $\theta_{(g,h)}$ is semiliftable is straightforward. We see that (θ,U) is the perturbation of the action $(g,h) \to \alpha_h^\omega$ by $(V_{gh^{-1}})_{(g,h)}$, and hence it is a cocycle crossed action. Let us show that $U_{(g,h),(k,\ell)} \in M_\omega$. For each $\phi \in M_*$ we have

$$\lim_{\nu\to\omega} \phi \circ \operatorname{Ad} v_g^\nu = \phi \circ \alpha_g \quad , \qquad g \in G \; .$$

If $u^\nu_{(g,h),(k,\ell)} = v^\nu_{gh^{-1}}\alpha_h(v^\nu_{k\ell^{-1}})v^{\nu *}_{gk\ell^{-1}h^{-1}}$, then

$$\lim_{\nu\to\omega} \phi \circ \operatorname{Ad} u^\nu_{(g,h),(k,\ell)} = \phi \circ (\alpha_{gh^{-1}}\alpha_h\alpha_{k\ell^{-1}}\alpha_{h^{-1}}\alpha^{-1}_{gk\ell^{-1}h^{-1}}) = \phi \; ,$$

hence $U_{(g,h),(k,\ell)} \in M_\omega$.

Let us show that for $(g,h) \neq (1,1)$, $\theta_{(g,h)}|M_\omega$ is strongly outer.

If $h \neq 1$, then for each ν, $\mathrm{Ad}\, v^{\nu}_{gh^{-1}}\alpha_h$ is centrally nontrivial and thus by Lemma 5.7, $\theta_{(g,h)}|M_\omega$ is strongly outer. If $h=1$ and $g \neq 1$, then $\lim_{\nu\to\omega} \mathrm{Ad}\, v^{\nu}_{gh^{-1}}\alpha_h = \alpha_g$ which is centrally nontrivial and Lemma 5.6 shows that $\theta_{(g,1)}$ is strongly outer. The lemma is proved.

9.2 We show that the approximate innerness of a group action, defined pointwise, can be given a global form.

LEMMA. *Let* $\alpha: G \to \mathrm{Aut}\, M$ *be a centrally free approximately inner action of an amenable group on a factor. There exist unitaries* V_g M^ω *represented by sequences* $(v_g)_\nu$ *of unitaries in* M, *with* $\lim_{\nu\to\omega} \mathrm{Ad}\, v^{\nu}_g = \alpha_g$ $V_1 = 1$ *and such that*

$$V_g V_h = V_{gh}$$

$$\alpha^\omega_g(V_h) = V_{ghg^{-1}} \qquad g,h \in G \ .$$

This can be restated as the fact that $(V_{gh^{-1}})_{(g,h)}$ is a cocycle for $(g,h) \mapsto (\alpha^\omega_h)$ and implies the fact that $\theta_{(g,h)} = \mathrm{Ad}\, V_{gh^{-1}}\alpha^\omega_h$ is an action.

<u>Proof</u>. Let $\bar{V}_g$, $g \in G$, be unitaries in M^ω, implementing (α_g) as in the previous lemma, and let $(\bar{\theta},\bar{U})$ be the corresponding cocycle crossed action of $G \times G$ on M^ω. Since $(\bar{\theta}|M_\omega, \bar{U})$ is strongly free and $G \times G$ is amenable, we can apply Proposition 7.4 to conclude that $\bar{U}$ is a coboundary, i.e. there exists a perturbation $(W_{(g,h)}) \subset M_\omega$ such that $(\bar{\theta},\bar{U})$ perturbed with W yields an action. This way $(W_{(g,h)}\bar{V}_{gh^{-1}})_{(g,h)}$ perturbs the action $(\alpha^\omega_h)_{(g,h)}$ to an action, and hence is a cocycle for it. Thus we have for $g,h,k,\ell \in G$,

$$(1) \quad W_{(g,h)}\bar{V}_{gh^{-1}}\alpha_h(W_{(k,\ell)}\bar{V}_{k\ell^{-1}}) = W_{(gk,h\ell)}\bar{V}_{gk\ell^{-1}h^{-1}} \ .$$

If we let $h=g$ and $\ell=k$ we obtain

$$W_{(g,g)}\alpha^\omega_g(W_{(k,k)}) = W_{(gk,gk)}$$

hence $(W_{(g,g)}) \subset M_\omega$ is a cocycle for the action $(\alpha_g)_\omega$ of G on M_ω which is strongly free. Proposition 7.2 yields a unitary $Z \in M_\omega$ with

$$W_{(g,g)} = Z^* \alpha^\omega_g(Z) \qquad g \in G \ .$$

We define

$$V_g = ZW_{(g,k)}\,\bar{V}_g Z^* \ .$$

Since V_g differs from $\bar{V}_g$ by unitaries in M_ω, V_g also implements α_g on M.

In (1), if we let $h = \ell = 1$ we infer

$$W_{(g,1)}\bar{V}_g W_{(k,1)}\bar{V}_k = W_{(gk,1)}\bar{V}_{gh}$$

which easily yields

$$V_g V_k = V_{gk} \quad .$$

In (1) we now substitute $hg^{-1}, 1, g, g$ for g,h,k,ℓ and obtain

$$W_{(hg^{-1},1)}\bar{V}_{gh^{-1}}W_{(g,g)} = W_{(h,g)}\bar{V}_{hg^{-1}}$$

and thus

$$\begin{aligned} V_{gh^{-1}} &= ZW_{(hg^{-1},1)}\bar{V}_{hg^{-1}}Z^* \\ &= ZW_{(h,g)}\bar{V}_{hg^{-1}}W^*_{(g,g)}Z^* \\ &= ZW_{(h,g)}\bar{V}_{hg^{-1}}\alpha^\omega_g(Z^*) \quad . \end{aligned}$$

In particular,

$$V_{g^{-1}} = ZW_{(1,g)}\,\bar{V}_{g^{-1}}\,\alpha^\omega_g(Z^*)$$

hence

$$\begin{aligned} V_{g^{-1}}\alpha^\omega_g(V_h) &= ZW_{(1,g)}\bar{V}_{g^{-1}}\alpha^\omega_g(Z^*ZW_{(h,1)}\bar{V}_hZ^*) \\ &= ZW_{(1,g)}\bar{V}_{g^{-1}}\alpha^\omega_g(W_{(h,1)}\,\bar{V}_h\,Z^*) \quad . \end{aligned}$$

If, in (1), we let $1,g,h,1$ stand for g,h,k,ℓ we get

$$W_{(1,g)}\bar{V}_{g^{-1}}\alpha_g(W_{(h,1)}\bar{V}_h) = W_{(h,g)}\bar{V}_{hg^{-1}}$$

which yields in the preceding equality

$$V_{g^{-1}}\alpha^\omega_g(V_h) = ZW_{(h,g)}\bar{V}_{hg^{-1}}\alpha^\omega_g(Z^*) = V_{hg^{-1}} \quad .$$

Since (V_g) was shown to be a representation,

$$\alpha^\omega_g(V_h) = V^*_{g^{-1}}V_{hg^{-1}} = V_{ghg^{-1}}$$

and the lemma is proved.

9.3 Let us recall Theorem 1.4 in a convenient form.

THEOREM. *Let* $\alpha: G \to \mathrm{Aut}\, M$ *be a centrally free and approximately inner action of an amenable group* G *on a McDuff factor* M. *Let* $\varepsilon > 0$, *let* F *be a finite subset of* G *and let* $\psi_0 \in M_*^+$. *Then there exists a cocycle* (v_g) *for* (α_g) *and a* II_1 *hyperfinite subfactor* $R \subset M$ *such that*

$$M = R \otimes (R' \cap M) \quad .$$

$(\mathrm{Ad}\, v_g\alpha_g)(R) = R$ *and* $(\mathrm{Ad}\, v_g\alpha_g | R)$ *is conjugate to the model action*

$$(\mathrm{Ad}\, v_g\alpha_g \mid R' \cap M) = \mathrm{id}_{R' \cap M}$$

$$\| v_g - 1 \|_{\psi_0}^{\#} \leqslant \varepsilon \, , \qquad g \in F \quad .$$

From the above statement we can easily obtain Theorem 1.4. Indeed, by Theorem 1.2 the model action $(\alpha_g^{(0)})$ is outer conjugate to $(\alpha_g^{(0)}) \otimes \mathrm{id}_R$; from the above theorem we infer that (α_g) is outer conjugate to $(\alpha_g^{(0)}) \otimes \mathrm{id}_{R' \cap M}$ and hence to $(\alpha_g^{(0)}) \otimes \mathrm{id}_R \otimes \mathrm{id}_{R' \cap M} = (\alpha_g^{(0)}) \otimes \mathrm{id}_M$; moreover, we have control over all cocycles that appear.

We obtain the copy of the model action in the theorem by applying successively the following lemma, which yields a copy of the submodel. Recall that $\varepsilon_n > 0$ and $G_n \subset\subset G$ are part of the Paving Structure 3.4. The sets $\bar{S}^n$ index the n-th finite dimensional submodel 4.5.

LEMMA. *In the conditions of the theorem, let* $n \geqslant 5$, *let* $p = |\bar{S}^n|$ *and let* Ψ, Ξ *and* Φ *be finite subsets of* M_*, Ψ *consisting of states*

There exists a II_1 *hyperfinite subfactor* e *of* M, *such that* $M = e \otimes (e' \cap M)$, *and a cocycle* $(\bar{v}_g)$ *for* (α_g), *such that letting* $(\bar{\alpha}_g) = (\mathrm{Ad}\, \bar{v}_g\alpha_g)$ *we have*

> $\bar{\alpha}_g(e) = e$ *and* $(\bar{\alpha}_g | e)$ *is conjugate to the submodel action and* $(\bar{\alpha}_g | e' \cap M)$ *is outer conjugate to* (α_g).

(1) $\| \bar{v}_g - 1 \|_{\psi}^{\#} \leqslant \varepsilon_{n-5} \qquad \psi \in \Psi \, , \quad g \in G_{n-4}$.

(2) $\| \xi \circ P_{e' \cap M} - \xi \| \leqslant 2p \sup_{g \in G_{n+1}} \xi \circ \alpha_g - \xi \qquad \xi \in \Xi$

(3) *For each* $\phi \in \Phi$ *there exists* $\eta_i \in e_*$, $\| \eta_i \| \leqslant 1$, $\phi_i \in (e' \cap M)_*$, $i = 1, 2, \ldots, p^2$ *such that*

$$\| \phi - \sum_i \eta_i \otimes \phi_i \| \leqslant \delta$$

$$\| \phi_i \circ (\alpha_g | e' \cap M) - \phi_i \| \leqslant \delta \quad , \quad i = 1, \ldots, p^2, \quad g \in G_{n+2} \; .$$

The proof of the lemma, which puts the whole machinery developed in this paper to work, will occupy the next section.

We remark that conditions (3) above express the fact that $\phi \approx \phi \circ \beta_g$, where $\beta_g = id_e \otimes (\bar{\alpha}_g | e' \cap M)$, i.e. most of the action $\bar{\alpha}_g$ is concentrated on e; its form allows us to work further in $e' \cap M$. On the other hand, if for some $\phi \in M_*$ we would have $\phi \approx \phi \circ P_{e' \cap M}$, then, since $\bar{\alpha}_g \approx \alpha_g$, $\bar{\alpha}_g | e$ is inner and $\beta_g \approx id$, we could infer

$$\phi \circ \alpha_g \approx \phi \circ P_{e' \cap M} \circ \bar{\alpha}_g \approx \phi \circ P_{e' \cap M} \circ \beta_g \approx \phi \circ \beta_g \approx \phi$$

and henceforth the form of condition (2) above.

Proof of the Theorem

Let $\bar{n} \geq 5$ be large enough to provide $3\varepsilon_{\bar{n}-5} < \varepsilon$ and $G_{\bar{n}-4} \supset F$. We let $p_n = |\bar{S}^n|$, $n \in N$. Let (Ψ_n), $n \geq \bar{n}-1$ be an ascending family of finite sets of states in M_*, with $\Psi_{\bar{n}-1} = \emptyset$, and $\bigcup_n \Psi_n$ total in M_*. We construct inductively for $n = \bar{n}, \bar{n}+1, \bar{n}+2, \ldots$ mutually commuting II_1 hyperfinite subfactors $\bar{e}^{\bar{n}}, \bar{e}^{\bar{n}+1}, \ldots$ of M, and cocycles $(v_g^{\bar{n}})$ for $(\alpha_g^{\bar{n}-1}) = (\alpha_g)$, $(v_g^{\bar{n}+1})$ for $(\alpha_g^{\bar{n}}) = (\mathrm{Ad}\, v_g^{\bar{n}}\, \alpha_g^{\bar{n}-1}), \ldots$, $(\bar{v}_g^{n+1})$ for $(\alpha_g^n) = (\mathrm{Ad}\, v_g^n \alpha_g^{n-1}), \ldots$ such that if e^n is the subfactor of M generated by $\bar{e}^{\bar{n}}, \bar{e}^{\bar{n}+1}, \ldots, \bar{e}^n$, with $e^{\bar{n}-1} = \mathbb{C}.1$, and if $v_g^n = \bar{v}_g^n \bar{v}_g^{n-1} \ldots \bar{v}_g^{\bar{n}}$ with $v_g^{\bar{n}-1} = 1$, then for each $n \geq \bar{n}$ we have

$$M = \bar{e}^n \otimes ((\bar{e}^n)' \cap M)$$

$\alpha_g^n(\bar{e}^n) = \bar{e}^n$ and $(\alpha_g^n | \bar{e}^n)$ is conjugate to the submodel action; $(\alpha_g^n (e^n)' \cap M)$ is outer conjugate to (α_g)

$$\bar{v}_g^n \in (e^{n-1})' \cap M$$

(4,n) $\| v_g^n - v_g^{n-1} \|^{\#}_{\psi_0} < 2\varepsilon_{n-5}$ $\qquad g \in G_{n-4}$

(5,n) $\| \psi \circ P_{(e^n)' \cap M} - \psi \| \leq 4\varepsilon_n$ $\qquad \psi \in \Psi_{n-1}$

(6,n) There exists $r_n \in N$ such that for any $\psi \in \Psi_n$ there are $\zeta_k \in (e^n)_*$, $\|\zeta_k\| \leq 1$, $\xi_k \in ((e^n)' \cap M)_*$, $k = 1, \ldots, r_n$ such that

$$\| \psi - \sum_k \zeta_k \otimes \xi_k \| \leq \varepsilon_{n+1}$$

$$\| \xi_k \circ (\alpha_g^n | (e^n)' \cap M) - \xi_k \| \leq p_{n+1}^{-1} r_n^{-1} \varepsilon_{n+1}, \quad g \in G_{n+2}.$$

Let $n \geq \bar{n}$ and suppose, if $n > \bar{n}$, that $\bar{e}^{\bar{n}}, \ldots, \bar{e}^{n-1}$ and $(\bar{v}_g^{\bar{n}}), \ldots, (\bar{v}_g^{n-1})$ satisfying the above conditions have been constructed.

Let $N = (e^{n-1})' \cap M$. Let $q \in \mathbb{N}$ be chosen such that the following condition holds:

(7) For any $\psi \in \Psi_n$ there are $\chi_i \in (e^{n-1})_*$, $\|\chi_i\| \leq 1$ and $\phi_i \in N_*$; $i = 1, \ldots, q$ with $\|\psi - \sum_i \chi_i \otimes \phi_i\| \leq \frac{1}{2}\varepsilon_{n+1}$.

We assume that in (6,n-1) and (7) above, a choice is made and kept fixed in all that follows for each ψ involved.

The action $(\alpha_g^{n-1}|N)$ is, by the induction hypothesis, outer conjugate to (α_g), and hence is centrally free and approximately inner, and N is a McDuff factor.

Let us apply the lemma in this section in order to obtain a cocycle $(\bar{v}_g^n) \subset N$ for (α_g^{n-1}) and a subfactor $\bar{e}^n$ of N such that letting $\alpha_g^n = \mathrm{Ad}\, \bar{v}_g^n \alpha_g^{n-1}$ the following assertions hold:

(8) $N = \bar{e}^n \otimes ((\bar{e}^n)' \cap N)$

$\alpha_g^n(\bar{e}^n) = \bar{e}^n$ and $(\alpha_g^n|\bar{e}^n)$ is conjugate to the submodel action and $(\alpha_g^n|(\bar{e}^n)' \cap N)$ is outer conjugate to $(\alpha_g^{n-1}|N)$

$\|\bar{v}_g^n - 1\|_{\psi_0}^{\#} \leq \varepsilon_{n-5}$

$\|\bar{v}_g^n - 1\|_{\psi_g}^{\#} \leq \varepsilon_{n-5}$

for $g \in G_{n-4}$, with $\psi_g = \psi_0 \circ \mathrm{Ad}\, v_g^{n-1}$

(9) For any $\psi \in \Psi_{n-1}$, if $\zeta_k \in (e^{n-1})_*$ and $\xi_k \in N_*$, $k = 1, \ldots, r_{n-1}$ were chosen in (6,n-1), then for $k = 1, \ldots, r_{n-1}$ we have

$$\|\xi_k \circ P_{(e^n)' \cap N} - \xi_k\| \leq 2p_n \sup_{g \in G_{n+1}} \|\xi_k \circ (\alpha_g^{n-1}|N) - \xi_k\| .$$

(10) For any $\psi \in \Psi_n$, with $\chi_i \in (e^{n-1})_*$ and $\phi_i \in N_*$, $i = 1, \ldots, q$ chosen in (7), there exist $\eta_{i,j} \in (\bar{e}^n)_*$, $\|\eta_{i,j}\| = 1$ and $\xi_{i,j} \in ((\bar{e}^n)' \cap N)_*$, $i = 1, \ldots, q$, $j = 1, \ldots, p_n^2$ with

$$\| \phi_i - \sum_j \eta_{i,j} \times \xi_{i,j} \| \leq \tfrac{1}{2} q^{-1} \varepsilon_{n+2}$$

$$\| \xi_{i,j} \circ (\alpha_g^n | (\bar{e}^n)' \cap N) - \xi_{i,j} \| \leq p_{n+1}^{-1} r_n^{-1} \varepsilon_{n+1} , \qquad g \in G_{n+2} .$$

From (8) we infer, in view of the inequality 7.7(1), if $g \in G_{n-4}$ and $\psi_g = \psi_0 \circ \mathrm{Ad}\, v_g^{n-1}$

$$\| v_g^n - v_g^{n-1} \|_{\psi_0}^{\#} = \| (\bar{v}_g^n - 1) v_g^{n-1} \|_{\psi_0}^{\#}$$

$$\leq 2^{\frac{1}{2}} (\| \bar{v}_g^n - 1 \|_{\psi_0}^{\#} + \| \bar{v}^n - 1 \|_{\psi_g}^{\#}) \leq 2 \varepsilon_{n-5}$$

and hence (4,n) is proved.

We have assumed that $\Psi_{\bar{n}-1} = \emptyset$, hence the statement (5,n) is void for $n = \bar{n}$. Suppose $n > \bar{n}$ and for $\psi \in \Psi_{n-1}$ let $\zeta_k \in (e^{n-1})_*$, $\zeta_k \in N_*$; $k = 1, \ldots, r_{n-1}$ be chosen in (6,n-1). With (9) we infer for each k,

$$\| \xi_k \circ P_{(\bar{e}^n)' \cap N} - \xi_k \| \leq 2 p_n \sup_{g \in G_{n+1}} \| \xi_k \circ (\alpha_g^{n-1} | N) - \xi_k \|$$

$$\leq 2 p_n p_n^{-1} r_{n-1}^{-1} \varepsilon_n = 2 r_{n-1}^{-1} \varepsilon_n$$

hence if $\bar{\psi} = \sum_k \zeta_k \otimes \xi_k \in M_*$, then

$$\| \psi \circ P_{(\bar{e}^n)' \cap M} - \psi \| \leq 2 \| \psi - \bar{\psi} \| + \| \bar{\psi} \circ P_{(\bar{e}^n)' \cap M} - \bar{\psi} \|$$

$$\leq 2 \varepsilon_n + \sum_k \| \zeta_k \| \, \| \xi_k \circ P_{(\bar{e}^n)' \cap M} - \xi_k \|$$

$$\leq 2 \varepsilon_n + r_{n-1} 2 r_{n-1}^{-1} \varepsilon_n = 4 \varepsilon_n$$

and this way (5,n) is proved.

Let $\psi \in \Psi_n$ and let $\chi_i \in (e^{n-1})_*$, $\phi_i \in N_*$, $i = 1, \ldots, q$ be chosen as in (7). Further on let $\eta_{i,j} \in (\bar{e}^n)_*$, $\xi_{i,j} \in (\bar{e}^n)' \cap N$, $i = 1, \ldots, q$, $j = 1, \ldots, p_n^2$ be chosen as in (10). We let

$$\zeta_{i,j} = \chi_i \otimes \eta_{i,j} \in (e^{n-1} \otimes \bar{e}^n)_* = (e^n)_*$$

and infer for any i,j

$$\| \zeta_{i,j} \| = \| \chi_i \| \, \| \eta_{i,j} \| \leq 1 ,$$

$$\| \psi - \sum_{i,j} \zeta_{i,j} \otimes \xi_{i,j} \| \leq \| \psi - \sum_i \chi_i \otimes \phi_i \| + \sum_i \| \chi_i \| \, \| \phi_i - \sum_j \zeta_{i,j} \otimes \xi_{i,j} \|$$

$$\leq \tfrac{1}{2} \varepsilon_{n+1} + q \cdot \tfrac{1}{2} q^{-1} \varepsilon_{n-1} = \varepsilon_{n+1} ,$$

$$\| \xi_{i,j} \circ (\alpha_g^n | (\bar{e}^n)' \cap N) - \xi_{i,j} \| \leq p_{n+1}^{-1} r_n^{-1} \varepsilon_{n+1} \quad , \qquad g \in G_{n+2}$$

and thus, if we reindex $(\zeta_{i,j}),(\xi_{i,j})$, $i=1,\dots,q$; $j=1,\dots,p_n^2$ with $k=1,\dots,r_n = qp_n^2$, we obtain (6,n) and end the proof of the induction step.

From (4,n) we infer, since $\sum_{n \geq \bar{n}} 2\varepsilon_{n-5} < 3\varepsilon_{\bar{n}-5} < \varepsilon$ and $\bigcup_n G_n = G$, that

$$v_g = \lim_{n \to \infty} v_g^n \quad \text{*-strongly}$$

exists for any $g \in G$ and yields an (α_g) cocycle which satisfies

$$\| v_g - 1 \|_{\psi_0}^{\#} < \varepsilon \qquad g \in F \subset G_{n-4} \quad .$$

We let R be the weak closure of $\bigcup_n e^n$ in M. The conditions (5,n) show, in view of Lemma 8.2, that R is a II_1 hyperfinite factor and $M = R \otimes (R' \cap M)$.

Let (β_g^n) be the action of G on M given by

$$\beta_g^n = \mathrm{id}_{e^n} \otimes (\alpha_g^n | (e^n)' \cap M) \qquad g \in G \quad .$$

For $\psi \in \Psi_n$ and $\zeta_k \in (e^n)_*$, $\xi_k \in ((e^n)' \cap M)_*$, $k=1,\dots,r_n$ chosen in (6,n), we let $\bar{\psi} = \sum_k \zeta_k \otimes \xi_k$ and infer

$$\begin{aligned} \| \psi \circ \beta_g^n - \psi \| &\leq 2 \| \psi - \bar{\psi} \| + \| \bar{\psi} \circ \beta_g^n - \bar{\psi} \| \\ &\leq 2\varepsilon_{n+1} + \sum_k \| \zeta_k \| \, \| \xi_k \circ (\alpha_g^n | (e^n)' \cap M) - \xi_k \| \\ &\leq 2\varepsilon_{n+1} + r_n p_{n+1}^{-1} r_n^{-1} \varepsilon_{n+1} \leq 3\varepsilon_{n+1} \end{aligned}$$

for $g \in G_{n+2}$. Since $\bigcup_n \Psi_n$ is total in M_* and $\bigcup_n G_n = G$, we obtain

$$\lim_{n \to \infty} \| \psi \circ \beta_g^n - \psi \| = 0 \qquad g \in G, \quad \psi \in M_* \quad .$$

Let $x \in R' \cap M$. For any $n \in \mathbb{N}$, $g \in G$

$$(\mathrm{Ad}\, v_g^n \alpha_g)(x) = \alpha_g^n(x) = \beta_g^n(x)$$

hence

$$\begin{aligned} (\mathrm{Ad}\, v_g \alpha_g)(x) &= w - \lim_{n \to \infty} (\mathrm{Ad}\, v_g^n \alpha_g)(x) \\ &= w - \lim_{n \to \infty} \beta_g^n(x) = x \end{aligned}$$

and thus $\mathrm{Ad}\, v_g \alpha_g | R' \cap M = \mathrm{id}_{R' \cap M}$. This ends the proof of the theorem.

9.4 In this last section we give the proof of the lemma stated in **9.3**. The first part of the proof will be similar to the one of the main lemma of the preceding chapter. In the second part we make use of the fact that the action is approximately inner, and hence implemented by unitaries in M^{ω}. We let the copy of the submodel that we construct almost contain these unitaries, and in this way concentrate on this copy of most of the action.

To simplify the notation, in what follows we denote the extension α_g^{ω} of α_g to M^{ω} by α_g. We recall that $\varepsilon_n > 0$, $G_n \subset\subset G$, the ε_n-paving subsets $(K_i^n)_{i \in I_n}$ of G and the approximate left g translations $\ell_g^n : \bigcup_i K_i^n \longrightarrow \bigcup_i K_i^n$ are part of the Paving Structure 3.4. The n-th finite dimensional submodel 4.5 had a s.m.u. indexed by $\bar{S}^n = \bigcup_i K_i^n \times S_i^n$. In view of the assumptions 3.5 we make use without further mention of the fact that ε_{k+1} is very small with respect to ε_k, for any $k \geq 0$.

<u>Step A</u>. We construct a s.m.u. $(\bar{E}_{s,t})$, $s,t \in \bar{S}^n$, replique of the n-th finite dimensional submodel in M, which is approximately equivariant for (α_g) and is fixed by $(\mathrm{Ad}\, V_g^* \alpha_g)$, where $V_g \in M^{\omega}$ are unitaries implementing α_g.

Let us begin by choosing, according to Lemma 9.2, unitaries $V_g \in M^{\omega}$, $g \in G$, $V_1 = 1$, which implement α_g on M, and such that

$$V_g V_h = V_{gh}$$

$$\alpha_g(V_h) = V_{ghg^{-1}} \qquad g,h \in G .$$

The action $(\mathrm{Ad}\, V_g^* \alpha_g) : G \to \mathrm{Aut}\, M$ will be denoted by $\mathrm{Ad}\, V^*\alpha$ and the action $(\mathrm{Ad}\, V_{gh^{-1}} \alpha_h) = (\mathrm{Ad}\, V_g\ \mathrm{Ad}\, V_h^* \alpha_h) : G \times G \to \mathrm{Aut}\, M^{\omega}$ will be denoted by $\mathrm{Ad}\, V \times \mathrm{Ad}\, V^*\alpha$. By Lemma 9.1, the restriction of this last action to M_{ω} is strongly free, and Lemma 8.3 shows that the fixed point algebra $(M\ )^{\mathrm{Ad}\, V \times \mathrm{Ad}\, V^*\alpha}$ is of the type II_1. We choose a s.m.u. $(F_{s,t})$, $s,t \in \bar{S}^n$ in $(M_{\omega})^{\mathrm{Ad}\, V \times \mathrm{Ad}\, V^*\alpha}$.

We now apply the Relative Rohlin Theorem 6.6 to obtain a partition of unity $(\bar{F}_{i,k})$, $i \in I_{n-1}$, $k \in K_i^{n-1}$ in $(M_{\omega})^{\mathrm{Ad}\, V^*\alpha}$, which is approximately equivariant for $(\alpha_g | (M_{\omega})^{\mathrm{Ad}\, V^*\alpha}) = (\mathrm{Ad}\, V_g | (M_{\omega})^{\mathrm{Ad}\, V^*\alpha})$: the estimates in 6.6 being better for small ε than those in the Rohlin Theorem 7.1, so we may suppose that $(\bar{F}_{i,k})$ satisfies the same requirements as its homonimous in Step A of **8.7**.

We proceed to define the almost equivariant s.m.u. $(\bar{E}_{s,t})$, $s,t \in \bar{S}^n$, out of $(F_{s,t})$ and $(\bar{F}_{i,k})$ by the same formulae as in **8.7**, Step A.

The s.m.u. $(\bar{E}_{s,t})$ thus defined will satisfy

$$|\alpha_g(\bar{E}_{(k_1,s_1),(k_2,s_2)}) - \bar{E}_{(gk_1,s_1),(gk_2,s_2)}|_{\tau} \leq 22\varepsilon_{n-1}^{\frac{1}{2}} |\bar{S}^n|^{-1}$$

for $g \in G_{n-1}$, $(k_1,s_1),(k_2,s_2) \in \tilde{S}^n$, where $\tilde{S}^n \subseteq \bar{S}^n$, with $|\tilde{S}^n| \geq (1-\varepsilon_n)|\bar{S}^n|$ as defined in 8.7, Step A. Moreover, in this case we have $(\bar{E}_{s,t}) \subset (M_\omega)^{\mathrm{Ad}\, V^*\alpha}$.

Step B. This step parallels Step B in **8.7**. We construct a unitary perturbation $(\bar{W}_g) \subset (M_\omega)^{\mathrm{Ad}\, V^*\alpha}$ for (α_g) such that if $(\bar{U}_g)$ are the approximate left g-translation unitaries associated to $(\bar{E}_{s,t})$,

$$\bar{U}_g = \sum_i \sum_{k,s} E_{(k_g,s),(k,s)}$$

with $i \in I_n$, $(k,s) \in K_i^n \times S_i^n$, $k_g = \ell_g^n(k)$, and $\bar{E} \subset M_\omega$ is the subfactor generated by $(\bar{E}_{s,t})$. Then

$$\mathrm{Ad}\, \bar{W}_g \alpha_g | \bar{E} = \mathrm{Ad}\, \bar{U}_g | \bar{E}$$

and

$$|\bar{W}_g - 1|_\tau \leq 90\varepsilon_{n-1}^{1/2}, \qquad g \in G_{n-1}.$$

Step C. With the Relative Rohlin Theorem instead of the Rohlin Theorem, we can repeat the proof of Proposition 7.4 to obtain the vanishing of the 2-cohomology of (α_g) in $(M_\omega)^{\mathrm{Ad}\, V^*\alpha}$, instead of M_ω. We may proceed as in Step C of **8.7** and construct a unitary perturbation $(\tilde{W}_g) \subset \bar{E}' \cap (M_\omega)^{\mathrm{Ad}\, V^*\alpha}$, such that if

$$W_g = \tilde{W}_g \bar{U}_g^* \bar{W}_g = \bar{U}_g^* \tilde{W}_g \bar{W}_g$$

then $(W_g) \subset (M_\omega)^{\mathrm{Ad}\, V^*\alpha}$ is an (α_g) cocycle, $\mathrm{Ad}\, W_g \alpha_g | \bar{E} = \mathrm{id}_{\bar{E}}$ and

(1) $\|\bar{U}_g W_g - 1\|_\tau < 7\varepsilon_{n-4}^{1/2}$, $\quad g \in G_{n-4}$.

Step D. The aim of this step is to replace the copy $((\bar{E}_{s,t}),(\bar{U}_g)) \subset M_\omega$ of the n-th finite dimensional submodel with a copy $((\tilde{E}_{s,t}),(\tilde{U}_g)) \subset M_\omega$, such that the unitaries $\tilde{U}_g$ are very close to the unitaries V_g implementing α_g, i.e. such that $\tilde{E}$ "contains" the action α_g.

The unitaries $(W_g) \in M_\omega$ defined in Step C formed an (α_g) cocycle fixed by $(\mathrm{Ad}\, V_g^* \alpha_g)$, hence they form an $(\mathrm{Ad}\, V_g)$ cocycle as well. Thus $(W_g V_g)$ is a representation of G into $(M^\omega)^{\mathrm{Ad}\, v^*\alpha}$ and $\mathrm{Ad}(W_g V_g)|\bar{E} = \mathrm{id}_{\bar{E}}$, $g \in G$.

We let $\tilde{V}_g = \bar{U}_g W_g V_g \in (M^\omega)^{\mathrm{Ad}\, V^*}$, and obtain $\mathrm{Ad}\, \tilde{V}_g|\bar{E} = \mathrm{Ad}\, \bar{U}_g|\bar{E}$.

We replace the partial isometries $\bar{E}_{s,t}$ in $\bar{E}$ by some partial isometries $\tilde{E}_{s,t}$ built from $\tilde{V}_g$ as follows. We begin by making a choice of an element $\hat{i} \in K_i^n$ for each $i \in I_n$. For each $i \in I_n$, $(k,s) \in K_i^n \times S_i^n$ and $h = k\hat{i}^{-1}$ we have, in view of the fact that $\hat{i} \in K_i^n$

and $h\hat{i} = k \in K_i^n$, $\ell_h^n(\hat{i}) = k$ and so

$$\mathrm{Ad}\,\tilde{V}_h(\bar{E}_{(\hat{i},s),(\hat{i},s)}) = \mathrm{Ad}\,\bar{U}_h(\bar{E}_{(\hat{i},s),(\hat{i},s)}) = \bar{E}_{(k,s),(k,s)} \cdot$$

Therefore the formulae

$$\tilde{E}_{(k,s),(m,t)} = \tilde{V}_h\bar{E}_{(\hat{i},s),(\hat{j},t)}\tilde{V}_\ell^*$$

$$= \bar{E}_{(k,s),(k,s)}\tilde{V}_h\tilde{V}_\ell^* = \tilde{V}_h\tilde{V}_\ell^*\bar{E}_{(m,t),(m,t)}$$

where $(k,s) \in K_i^n \times S_i^n$, $(m,t) \in K_j^n \times S_j^n$, $h = k\hat{i}^{-1}$, $\ell = m\hat{j}^{-1}$, define a s.m.u. in $(M^\omega)^{\mathrm{Ad}\,V^*}$ with the same diagonal m.a.s.a. as $\bar{E}$, i.e.

$$\tilde{E}_{s,s} = \bar{E}_{s,s} \in M_\omega \qquad s \in \bar{S}^n \quad .$$

Let $\tilde{U}_g$ be the left g-translation unitary associated to $(\tilde{E}_{s,t})$

$$\tilde{U}_g = \sum_i \sum_{k,s} \tilde{E}_{(k_g,s),(k,s)} \qquad g \in G$$

with $i \in I_n$, $(k,s) \in K_i^n \times S_i^n$, $k_g = \ell_g^n(k)$.

Since $(\bar{U}_g^*\tilde{V}_g) = (W_gV_g)$ is a representation,

$$\bar{U}_{gh}^*\tilde{V}_{gh} = \bar{U}_g^*\tilde{V}_g\bar{U}_h^*\tilde{V}_h \qquad g,h \in G$$

and in view of the fact that $\mathrm{Ad}(\bar{U}_g^*\tilde{V}_g)|\bar{E} = \mathrm{Ad}(W_gV_g)|\bar{E} = \mathrm{id}_{\bar{E}}$ and $\bar{U}_g \in \bar{E}$ we infer

$$(2) \qquad \tilde{V}_{gh}\tilde{V}_h\tilde{V}_g^* = \bar{U}_{gh}(\bar{U}_g^*\tilde{V}_g)\bar{U}_h^*(\tilde{V}_g^*\bar{U}_g)\bar{U}_g^* = \bar{U}_{gh}\bar{U}_g^*\bar{U}_h^* \quad .$$

Let us keep $g \in G_n$ fixed. We have

$$\tilde{U}_g\tilde{V}_g^* - 1 = \sum_i \sum_{k,s} \bar{E}_{(k_1,s),(k_1,s)}(\tilde{V}_{h_1}\tilde{V}_h^*\tilde{V}_g^* - 1) = \Sigma_1 + \Sigma_2$$

where $i \in I_n$, $(k,s) \in K_i^n \times S_i^n$, $k_1 = \ell_g^n(k)$, $h = k\hat{i}^{-1}$, $h_1 = k_1\hat{i}^{-1}$; in Σ_1 we sum those terms in which $gk \in K_i^n$ and in Σ_2 those in which $k \in K_i^n \setminus g^{-1}K_i^n$.

Let $\psi \in M_*$ be a state. We have

$$|\Sigma_2|_\psi \leq \sum_i \sum_{k,s} |\bar{E}_{(k_1,s),(k_1,s)}|_\psi \, \|\tilde{V}_{h_1}\tilde{V}_h^*\tilde{V}_g^* - 1\|$$

$$\leq 2\sum_i \sum_{k,s} |\bar{E}_{(k_1,s),(k_1,s)}|_\tau$$

$$= 2\sum_i |K_i^n \setminus g^{-1}K_i^n|\,|S_i^n|\,|\bar{S}^n|^{-1}$$

$$\leq 2\varepsilon_n \sum_i |K_i^n|\,|S_i^n|\,|\bar{S}^n|^{-1} = 2\varepsilon_n$$

where $i \in I_n$, $k \in K_i^n \setminus g^{-1} K_i^n$, $s \in S_i^n$, and we have used the fact that K_i^n is (ε_n, G_n) invariant for the estimates of Lemma 7.1.

On the other hand, for a term in Σ_1 we have $k_1 = \ell_g^n(k) = gk$, $h = k\hat{i}^{-1}$ and $h_1 = gk\hat{i}^{-1}$ so that with (2) we infer

$$\begin{aligned} E_{(k_1,s),(k_1,s)} \tilde{V}_{h_1} \tilde{V}_h^* \tilde{V}_g^* &= \bar{E}_{(gk,s),(gk,s)} \bar{U}_{gk\hat{i}^{-1}} \bar{U}_{k\hat{i}^{-1}}^* \bar{U}_g^* \\ &= \bar{E}_{(gk,s),(\hat{i},s)} \bar{U}_{k\hat{i}^{-1}}^* \bar{U}_g^* \\ &= \bar{E}_{(gk,s),(k,s)} \bar{U}_g^* \\ &= \bar{E}_{(gk,s),(gk,s)} \end{aligned}$$

and thus $\Sigma_1 = 0$.

In conclusion, for any state $\psi \in M_*$ and $g \in G_n$,

$$|\tilde{U}_g \tilde{V}_g^* - 1|_\psi \leqslant |\Sigma_1|_\psi + |\Sigma_2|_\psi \leqslant 2\varepsilon_n \quad .$$

Analogue estimates yield

$$|\tilde{V}_g \tilde{U}_g^* - 1|_\psi \leqslant 2\varepsilon_n$$

hence with the inequality 7.1(7) we infer

$$\| \tilde{U}_g \tilde{V}_g^* - 1 \|_\psi^\# \leqslant 2\varepsilon_n^{\frac{1}{2}} \quad .$$

From this together with (1), we obtain for $g \in G_{n-4}$

$$\begin{aligned} (3) \qquad \| \tilde{U}_g V_g^* - 1 \|_\psi^\# &\leqslant 2^{\frac{1}{2}} (\| \tilde{U}_g \tilde{V}_g^* - 1 \|_\psi^\# + \| \tilde{V}_g V_g^* - 1 \|_\psi^\#) \\ &= 2^{\frac{1}{2}} (\| \tilde{U}_g \tilde{V}_g^* - 1 \|_\psi^\# + \| W_g V_g - 1 \|_\tau^\#) \\ &\leqslant 2^{\frac{1}{2}} (2\varepsilon_n + 6\varepsilon_{n-4}^{\frac{1}{2}}) < 10\varepsilon_{n-4}^{\frac{1}{2}} \end{aligned}$$

for any state $\psi \in M_*$.

<u>Step E</u>. We lift the whole construction done before from M^ω to M. For $g, h \in G$ we have

$$V_g^* \alpha_g (V_h^*) = V_g^* \mathrm{Ad}\, V_g (V_h^*) = V_{gh}^*$$

hence (V_g^*) is an (α_g) cocycle; moreover, $(\tilde{E}_{s,t}) \subset (M^\omega)^{\mathrm{Ad}\, V^* \alpha}$. We apply Lemma 8.4 and obtain representing sequences $(\tilde{e}_{s,t}^\nu)_\nu$ consisting of s.m.u.'s in M for $(\tilde{E}_{s,t})$ and representing sequences $(v_g^{\nu *})$ consisting of (α_g) cocycles in M for (V_g^*) such that if $\tilde{e}^\nu$ is the subfactor of M generated by $(\tilde{e}_{s,t}^\nu)$ then

$$\mathrm{Ad}\, v_g^{\nu *}\alpha_g | \tilde{e}^\nu = \mathrm{id} \qquad g \in G, \quad \nu \in \mathbb{N} .$$

For $g \in G$, we define with the usual formulae the unitary $\tilde{u}_g^\nu \in M$ associated to $(\tilde{e}_{s,t}^\nu)$ such that $((\tilde{e}_{s,t}^\nu),(\tilde{u}_g^\nu))$ is a copy of the n-th finite dimensional submodel. Then $(\tilde{U}_g^\nu)_\nu$ will represent $\tilde{U}_g \in M^\omega$. We let $\tilde{\alpha}_g^\nu = \mathrm{Ad}\, v_g^{\nu *}\alpha_g \in \mathrm{Aut}\, M$. From (3) we have, for any state $\psi \in M_*$

$$(6) \qquad \lim_{\nu \to \omega} \|\tilde{u}_g^\nu v_g^{\nu *} - 1\|_\psi^\# < 9\varepsilon_{n-4}^{1/2} , \qquad g \in G_{n-4}$$

and also, since $\tilde{E}_{s,s} = \bar{E}_{s,s} \in M_\omega$

$$(7) \qquad \lim_{\nu \to \omega} (\tilde{e}_{s,s}^\nu) = \psi^\omega(\tilde{E}_{s,s}) = \psi_\omega(\bar{E}_{s,s}) = \tau(\bar{E}_{s,s}) = |\bar{S}^n|^{-1} \quad s \in \bar{S}^n .$$

We now study the decomposition of M_* with respect to $M = \tilde{e}^\nu \otimes ((\tilde{e}^\nu)' \cap M)$. Let $\tilde{\eta}_{s,t}^\nu \in (\tilde{e}^\nu)_*$ be the basis dual to $\tilde{e}_{s,t}$, $s,t \in \bar{S}^n$. For $\phi \in M_*$ let

$$\phi = \sum_{s,t} \tilde{\eta}_{s,t}^\nu \otimes \tilde{\phi}_{s,t}^\nu$$

with $\tilde{\phi}_{s,t}^\nu = (\tilde{e}_{s,t}^\nu \phi) | (\tilde{e}^\nu)' \cap M$. We have $\tilde{\alpha}_g^\nu | \tilde{e}^\nu = \mathrm{id}$, hence for any $x \in (\tilde{e}^\nu)' \cap M$ and $s,t \in \bar{S}^n$,

$$\begin{aligned} |\tilde{\phi}_{s,t}^\nu(\tilde{\alpha}_g^\nu(x) - x)| &= |\phi((\tilde{\alpha}_g^\nu(x) - x)\tilde{e}_{s,t}^\nu)| \\ &= |\phi(\tilde{\alpha}_g^\nu(x\tilde{e}_{s,t}^\nu) - x\tilde{e}_{s,t}^\nu)| \\ &\leq \|\phi \circ \tilde{\alpha}_g^\nu - \phi\| \ \|x\| . \end{aligned}$$

Since $\mathrm{Ad}\, v_g^\nu \to \alpha_g$ when $\nu \to \omega$, we have $\alpha_g^\nu \to \mathrm{id}$ and so

$$(8) \qquad \lim_{\nu \to \omega} \|\tilde{\phi}_{s,t}^\nu \circ \tilde{\alpha}_g^\nu - \tilde{\phi}_{s,t}^\nu\| = 0$$

for any $\phi \in M_*$, $s,t \in \bar{S}^n$ and $g \in G$.

We now show that if a state is almost invariant with respect to (α_g) then it almost commutes with $(\tilde{e}_{s,t}^\nu)$.

Let $i,j \in I_n$, $\bar{s} = (k,s) \in K_i^n \times S_i^n$, $\bar{t} = (m,t) \in K_j^n \times S_j^n$, $h = k\hat{\imath}^{-1}$, $\ell = m\hat{\jmath}^{-1}$ where $\hat{\imath} \in K_i^n$ and $\hat{\jmath} \in K_j^n$ were defined in Step D. We have

$$\begin{aligned} \tilde{E}_{\bar{s},\bar{t}} V_{h\ell^{-1}}^* &= \bar{E}_{\bar{s},\bar{s}} \tilde{V}_h \tilde{V}^* V_{h\ell^{-1}}^* = \bar{E}_{\bar{s},\bar{s}} \bar{U}_h^* W_h V^* W^* U_\ell V_{h\ell^{-1}}^* \\ &= \bar{E}_{\bar{s},\bar{s}} U_h^* W_h \ \mathrm{Ad}\, V_{h\ell^{-1}}(W_\ell^* \bar{U}_\ell) \in M_\omega \end{aligned}$$

and since, by assumptions 3.5,

$$h\ell^{-1} \in K_i^n(K_i^n)^{-1}\ K_j^n(K_j^n)^{-1} \subset G_{n+1}$$

we conclude that for any $s,t \in \bar{S}^n$ there exists $g \in G_{n+1}$ with $\tilde{E}_{s,t}V_g^* \in M_\omega$.

Let $\xi \in M_*$. We have

$$\lim_{\nu\to\omega} \|\xi \circ P_{(\tilde{e}^\nu)' \cap M} - \xi\| = \lim_{\nu\to\omega} \Big\| |\bar{S}^n|^{-1} \big(\sum_{s,t} \tilde{e}^\nu_{s,t}\xi\tilde{e}^\nu_{t,s} - \tilde{e}^\nu_{s,t}\tilde{e}^\nu_{t,s}\xi \big) \Big\|$$

$$\leqslant |\bar{S}^n|^{-1} \sum_{s,t} \lim_{\nu\to\omega} \| [\tilde{e}^\nu_{s,t}, \xi] \| .$$

For $s,t \in \bar{S}^n$, let $g \in G_{n+1}$ be such that $E_{s,t}V_g^* \in M_\omega$, as yielded by the previous discussion and let $(x^\nu)_\nu = (\tilde{e}^\nu_{s,t}v_g^{\nu *})$ be the corresponding ω-centralizing sequence. We infer

$$\lim_{\nu\to\omega} \| [\tilde{e}^\nu_{s,t}, \xi] \| = \lim_{\nu\to\omega} \| x^\nu v_g^\nu \xi - \xi x^\nu v_g^\nu \|$$

$$\leqslant \lim_{\nu\to\omega} (\| x^\nu v_g^\nu (\xi - \xi \circ \mathrm{Ad}\, v_g^\nu) \| + \| [x^\nu, \xi] v_g \|)$$

$$\leqslant \lim_{\nu\to\omega} (\| \xi - \xi \circ \mathrm{Ad}\, v_g^\nu \| + \| [x^\nu, \xi] \|)$$

$$= \| \xi - \xi \circ \alpha_g \| .$$

In conclusion, for any $\xi \in M_*$ we obtain

$$(9) \qquad \lim_{\nu\to\omega} \|\xi \circ P_{(e^\nu)' \cap M} - \xi\| \leqslant |\bar{S}^n| \sup_{g \in G_{n+1}} \|\xi - \xi \circ \alpha_g\| .$$

An appropriate choice of ν yields, in view of (6)-(9) above, a s.m.u. $(\tilde{e}_{s,t})$, $s,t \in \bar{S}^n$ in M and a cocycle (v_g^*) for (α_g) such that if $\tilde{e}$ is the subfactor of M generated by $(\tilde{e}_{s,t})$ and if $\tilde{\alpha}_g = \mathrm{Ad}\, v_g^* \alpha_g$ then there are unitaries $\tilde{u}_g \in \tilde{e}$, $g \in G$ such that $((\tilde{e}_{s,t}),(\tilde{u}_g))$ is a copy of the n-th finite dimensional submodel 4.5 $\tilde{\alpha}_g|\tilde{e} = \mathrm{id}$, $g \in G$.

$$(10) \qquad \|\tilde{u}_g v_g^* - 1\|_\psi^\# \leqslant 10\varepsilon_{n-4}^{1/2} \qquad g \in G_{n-4}, \quad \psi \in \Psi$$

$$(11) \qquad \psi(\tilde{e}_{s,s}) \leqslant (1+\varepsilon_n)|\bar{S}^n|^{-1} \qquad s \in \bar{S}^n, \quad \psi \in \Psi$$

$$(12) \qquad \|\xi \circ P_{\tilde{e}' \cap M} - \xi\| \leqslant 3/2\, |\bar{S}^n| \sup_{g \in G_{n+1}} \|\xi \circ \alpha_g - \xi\| , \qquad \xi \in \Xi$$

$$(13) \qquad \|\tilde{\phi}_{s,t} \circ \tilde{\alpha}_g - \tilde{\phi}_{s,t}\| \leqslant 1/4\, \delta \qquad \phi \in \Phi, \quad s,t \in \bar{S}^n, \quad g \in G_{n+2}$$

where $\phi = \sum_{s,t} \tilde{\eta}_{s,t} \otimes \tilde{\phi}_{s,t}$, with $(\tilde{\eta}_{s,t}) \subset \tilde{e}_*$ the basis dual to $(\tilde{e}_{s,t})$ and $\tilde{\phi}_{s,t} = (\tilde{e}_{s,t}\phi)|\tilde{e}' \cap M$.

Step F. We complete the copy $\tilde{e}$ of the finite dimensional submodel with a subfactor f, almost fixed by $(\tilde{\alpha}_g)$ and thus obtain a copy of the submodel.

Let $N = \tilde{e}' \cap M$. Then by **5.8**, N is McDuff and $(\tilde{\alpha}_g|N)$ is centrally free. We apply Theorem 8.5 to construct a II_1 hyperfinite subfactor $f \subset N$ with $N = f \otimes (f' \cap N)$ and a cocycle (z_g) for $(\tilde{\alpha}_g)$ such that

$$(\mathrm{Ad}\, z_g\tilde{\alpha}_g|f) = \mathrm{id}_f$$

$$(\mathrm{Ad}\, z_g\tilde{\alpha}_g|f' \cap N) \text{ is outer conjugate to } \mathrm{id}_{\tilde{e}} \times (\alpha_g|f' \cap N) = (\tilde{\alpha}_g)$$

$$(14) \quad \|z_g - 1\|_\psi^\# \leq \varepsilon_n \qquad g \in G_{n-4}, \quad \psi \in \Psi$$

$$(15) \quad \|(\xi|N) \circ P_{f' \cap N} - \xi|N\| \leq \sup_{g \in G_{n+1}} \|\xi \circ \alpha_g - \xi\|, \quad \xi \in \Xi$$

and with $\tilde{\psi}_{s,t} = (\tilde{e}_{s,t}\psi)|N$, $\psi \in M_*$, $s,t \in \bar{S}^n$,

$$(16) \quad \|\tilde{\psi}_{s,s} \circ P_{f' \cap N} - \tilde{\psi}_{s,s}\| \leq \varepsilon_n |\bar{S}^n|^{-1} \qquad \psi \in \Psi, \quad s \in \bar{S}^n$$

$$(17) \quad \|\tilde{\phi}_{s,t} \circ P_{f' \cap N} - \tilde{\phi}_{s,t}\| \leq \tfrac{1}{4}|\bar{S}^n|^{-2} \qquad \psi \in \Psi, \quad s,t \in \bar{S}^n .$$

We extend the isomorphism between e and the finite dimensional submodel to an isomorphism between $e = \tilde{e} \otimes f$ and the submodel. Let τ be the normalized trace on e and let $(u_g) \subset e$ be the copy of the submodel representation. Let a be the m.a.s.a. of e which is the copy of the diagonal m.a.s.a. of the submodel. Then a is generated by $(\tilde{e}_{s,s})$, $s \in \bar{S}^n$ and by a m.a.s.a. of the subfactor f. It is now that we use the estimates in Corollary 4.4 instead of Lemma 4.4, the reason being that only the diagonal m.a.s.a. of $\tilde{e}$ comes from M_ω and thus behaves well with respect to M_*.

For each $g \in G_n$ there exists a projection $p_g \in a$, with $\tau(p_g) \leq 8\varepsilon_n$, such that $(1 - p_g)u_g = (1 - p_g)\tilde{u}_g$.

Let $p_g = \sum_s \tilde{e}_{s,s}p_{g,s}$, $s \in \bar{S}^n$ with $p_{g,s}$ projections in f. Then

$$\tau(p_g) = \sum_s \tau(\tilde{e}_{s,s})\tau(p_{g,s}) = |\bar{S}^n|^{-1} \sum_s \tau(p_{g,s}) .$$

For $\psi \in \Psi$, $s \in \bar{S}^n$ and $g \in G_n$ we have, in view of (11) and (17),

$$\begin{aligned}
\psi(\tilde{e}_{s,s}p_{g,s}) &= \tilde{\psi}_{s,s}(p_{g,s}) \\
&\leq \tilde{\psi}_{s,s}(P_{f'\cap N}(p_{g,s})) + \|\tilde{\psi}_{s,s}\circ P_{g'\cap N} - \tilde{\psi}_{s,s}\| \\
&\leq \tilde{\psi}_{s,s}(\tau(p_{g,s})) + \varepsilon_n|\bar{S}^n|^{-1} \\
&= \psi(\tilde{e}_{s,s})\tau(p_{g,s}) + \varepsilon_n|\bar{S}^n|^{-1} \\
&\leq (1+\varepsilon_n)|\bar{S}^n|^{-1}\tau(p_{g,s}) + \varepsilon_n|\bar{S}^n|^{-1} \quad .
\end{aligned}$$

Hence

$$\begin{aligned}
\psi(p_g) &= \sum_s \psi(\tilde{e}_{s,s}p_{g,s}) \leq (1+\varepsilon_n)|\bar{S}^n|^{-1}\sum_s \tau(p_{g,s}) + \varepsilon_n \\
&= (1+\varepsilon_n)\tau(p_g) + \varepsilon_n \leq (1+\varepsilon_n)8\varepsilon_n + \varepsilon_n < 10\varepsilon_n \quad .
\end{aligned}$$

We infer

$$\begin{aligned}
(19) \quad \|u_g\tilde{u}_g^* - 1\|_\psi^\# &= \|p_g(u_g\tilde{u}_g^* - 1)p_g\|_\psi^\# \\
&\leq \|u_g\tilde{u}_g^* - 1\| \ \psi(p_g)^{\frac{1}{2}} \leq 2\cdot(10\varepsilon_n)^{\frac{1}{2}} < 7\varepsilon_n^{\frac{1}{2}} \quad .
\end{aligned}$$

We have $\mathrm{Ad}\, v_g^*\alpha_g|\tilde{e} = \mathrm{id}$, $z_g \in \tilde{e}'\cap M$ and $\mathrm{Ad}(z_g v_g^*)\alpha_g|f = \mathrm{id}$, hence $\mathrm{Ad}(z_g v_g^*)\alpha_g|e = \mathrm{id}$.

Since (u_g) is a representation of G in e, (u_g) is a cocycle for $\mathrm{Ad}(z_g v_g^*)\alpha_g$, thus if we let $\bar{v}_g = u_g z_g v_g^*$ then $(\bar{v}_g)$ is a cocycle for (α_g), and $\mathrm{Ad}\,\bar{v}_g\alpha_g|e = \mathrm{Ad}\, u_g|e$, i.e. $(\mathrm{Ad}\,\bar{v}_g\alpha_g|e)$ is conjugate to the submodel action.

As $\tilde{u}_g \in \tilde{e}$ and $z_g \in \tilde{e}'\cap M$, we have, via 7.1(10),

$$\begin{aligned}
\|v_g - 1\|_\psi^\# &= \|u_g\tilde{u}_g^* z_g\tilde{u}_g v_g^* - 1\|_\psi^\# \\
&\leq 2(\|u_g\tilde{u}_g^* - 1\|_\psi^\# + \|z_g - 1\|_\psi^\# + \|\tilde{u}_g v_g^* - 1\|_\psi^\#) \\
&\leq 2(7\varepsilon_n^{\frac{1}{2}} + \varepsilon_n + 10\varepsilon_{n-4}^{\frac{1}{2}}) < \varepsilon_{n-5}
\end{aligned}$$

for $g \in G_{n-4}$, $\psi \in \Psi$, where we have used (19), (14) and (10) before. The estimate (1) in Lemma 9.3 is now proved.

From (12) and (15) for $\xi \in \Xi$ we have

$$\begin{aligned}
\|\xi\circ P_{e'\cap M} - \xi\| &\leq \|\xi\circ P_{e'\cap M} - \xi\| + \|(\xi|N)\circ P_{f'\cap N} - (\xi|N)\| \\
&\leq (|\bar{S}^n| + 1) \sup_{g\in G_{n+1}} \|\xi\circ\alpha_g - \xi\|
\end{aligned}$$

which proves statement (2) in the lemma.

Let σ be the normalized trace of f. For $\phi \in \Phi$ with

$$\phi = \sum_{s,t} \eta_{s,t} \otimes \phi_{s,t} \in (\tilde{e} \otimes N)_* = M_*$$

as in (13) and (17), we let

$$\eta_{s,t} = \tilde{\eta}_{s,t} \otimes \sigma \in (\tilde{e} \otimes f)_* = e_*$$

$$\phi_{s,t} = \tilde{\phi}_{s,t}|e' \cap M .$$

We have $\tilde{\phi}_{s,t} \circ P_{f' \cap N} = \sigma \otimes \phi_{s,t}$, hence

$$\begin{aligned} \|\phi \sum_{s,t} \eta_{s,t} \otimes \phi_{s,t}\| &= \| \sum_{s,t} \tilde{\eta}_{s,t} \otimes (\tilde{\phi}_{s,t} - \tilde{\phi}_{s,t} \circ P_{f' \cap N}) \| \\ &\leqslant \sum_{s,t} \|\tilde{\phi}_{s,t} - \tilde{\phi}_{s,t} \circ P_{f' \cap N}\| \\ &\leqslant \tfrac{1}{4}|\bar{S}^n|^2 \, |\bar{S}^n|^{-2}\delta = \tfrac{1}{4}\delta \quad . \end{aligned}$$

Since

$$\bar{\alpha}_g|e' \cap M = \mathrm{Ad}(u_g z_g v_g^*)\alpha_g|e' \cap M = \mathrm{Ad}(z_g v_g^*)\alpha_g|e' \cap M$$

and

$$\mathrm{Ad}(z_g v_g^*)\alpha_g|e = \mathrm{id}$$

we infer

$$\begin{aligned} &\|\phi_{s,t} \circ (\bar{\alpha}_g|e' \cap M) - \phi_{s,t}\| \\ &\quad = \|(\sigma \otimes \phi_{s,t}) \circ (\mathrm{Ad}(z_g v_g^*)\alpha_g|e' \cap M) - \sigma \otimes \phi_{s,t}\| \\ &\quad \leqslant 2\|\tilde{\phi}_{s,t} P_{f' \cap N} - \tilde{\phi}_{s,t}\| + \|\tilde{\phi}_{s,t} \circ \mathrm{Ad}\, z_g - \tilde{\phi}_{s,t}\| \\ &\qquad + \|\tilde{\phi}_{s,t} \circ (\mathrm{Ad}\, v_g \alpha_g|\tilde{e}' \cap M) - \tilde{\phi}_{s,t}\| \\ &\quad < 2 \cdot \tfrac{1}{4}|\bar{S}^n|^{-2}\delta + \tfrac{1}{4}\delta + \tfrac{1}{4}\delta \leqslant \delta \end{aligned}$$

where we have used (17), (18) and (19).

The last estimate in the lemma is thus obtained by reindexing $\eta_{s,t}, \phi_{s,t}$, $s,t \in |\bar{S}^n|$ with a single index $k = 1,2,\dots,|\bar{S}^n|^2 = p^2$.

Which brings us to the END.

REFERENCES

[1] A.CONNES: "Almost periodic states and factors of type III_1," J. Funct. Anal., **16** (1974), 415-445.

[2] A.CONNES: "Une classification des facteurs de type III," Ann. Sci. Ec. Norm. Sup. 4me serie, t. 6, fasc. 2 (1973), 133-152.

[3] A.CONNES: "Periodic automorphisms of the hyperfinite factor of type II_1," Acta Sci. Math. **39** (1977), 39-66.

[4] A.CONNES: "Outer conjugacy classes of automorphisms of factors," Ann. Sci. Ec. Norm. Sup. 4me serie, t. 8 (1975),, 383-420.

[5] A.CONNES: "Classification of injective factors," Ann. of Math. **104** (1976), 73-115.

[6] A.CONNES: "On the classification of von Neumann algebras and their automorphisms," Symposia Math. XX (1976), 435-578.

[7] A.CONNES and M.TAKESAKI: "The flow of weights on factors of type III," Tohoku Math. J. **29** (1977), 473-575.

[8] E.B.DAVIES: "Involutory automorphisms of operator algebras," Trans. A.M.S. **158** (1971), 115-142.

[9] J.DIXMIER: Les Algebres d'Operateurs Dans L'espace Hilbertien, Deuxieme edition, (Gauthier Villars, 1969).

[10] S.EILENBERG and S.MacLANE: "Cohomology theory in abstract groups, II," Ann. of Math. **48** (1947), 326-341.

[11] T.FACK and O.MARECHAL: "Sur la classification des automorphismes périodiques de C^*-algebres U.H.F.," Canadian Jour. of Math. **31**, no. 3 (1979), 469-523.

[12] E.FØLNER: "On groups with full Banach mean value," Math. Scand. **3** (1955), 243-254.

[13] T.GIORDANO: Antiautomorphismes involutifs des facteurs de von Neumann injectifs. These, Neuchatel.

[14] T.GIORDANO and V.JONES: "Antiautomorphismes involutifs du facteur hyperfini de type II_1," C. R. Acad. Sci. Paris, s.A **290** (1980), 29-31.

[15] F.P.GREENLEAF: Invariant Means on Topological Groups, Van Nostrand Math. Studies No. 16.

[16] R.I.GRIGORCHUK: "Symmetrical random walks on discrete groups," Uspehi Math. Nauk. **32** (1977), 217-218.

[17] R.HERMAN and V.F.R.JONES: "Period two automorphisms of UHF C^*-algebras," to appear J. Funct. Analysis.

[18] R.HERMAN and V.F.R.JONES: "Models of finite group actions," J. Funct. Anal. **45** (1982).

[19] J.HUEBSCHMANN: Dissertation, E.T.H., Zürich.

[20] V.F.R.JONES: "Sur la conjugaison de sous-facteurs de type II_1," C. R. Acad. Sci. Paris, **284** (1977), 597-598.

[21] V.F.R.JONES: "An invariant for group actions," in Algèbres d'Opérateurs, Springer Lecture Notes in Mathematics, No. 725.

[22] V.F.R.JONES: "Actions of finite abelian groups on the hyperfinite II_1 factor," preprint.

[23] V.F.R.JONES: "Actions of finite groups on the hyperfinite type II factor," Memoirs A.M.S. No. 237 (1980).

[24] V.F.R.JONES: "The spectrum of a finite group action," preprint.

[25] V.F.R.JONES: "Prime actions of compact abelian groups on the hyperfinite II_1 factor," J. Operator Theory **9** (1983), 181-186.

[26] V.F.R.JONES: "A converse to Ocneanu's theorem," J. Operator Theory **10** (1983), 61-63.

[27] V.F.R.JONES and S.POPA: "Some properties of MASA's in factors," Operator Theory: Adv. Appl. **6**, (Birkhäuser, 1982).

[28] A.KISHIMOTO: "On the fixed point algebra of a UHF algebra under a periodic automorphism of product type," Publ. RIMS Kyoto University **13** (1977), 777-791.

[29] W.KRIEGER: "On ergodic flows and the isomorphism of factors," Math. Ann. **223** (1976), 19-70.

[30] D.McDUFF: "Central sequences and the hyperfinite factor," Proc. London Math. Soc. **XXI** (1970), 443-461.

[31] F.J.MURRAY and J. von NEUMANN: "Rings of operators, IV," Ann. of Math. **44** (1943), 716-808.

[32] M.NAKAMURA and Z.TAKEDA: "On the extensions of finite factors, II," Proc. Jap. Acad. **35** (1959), 215-220.

[33] A.OCNEANU: "A Rohlin theorem for groups acting on von Neumann algebras," in Topics in Modern Operator Algebra, (Birkhäuser Verlag, 1981), 247-258.

[34] A.OCNEANU: "Actions des groupes moyennables sur les algebres de von Neumann," C. R. Acad. Sci. Paris, **291** (1980), 399-401.

[35] A.OCNEANU: "Actions of compact groups on factors," in preparation.

[36] D.ORNSTEIN and B.WEISS: "Ergodic theory of amenable group actions, I," Bull. A.M.S., vol. **2**, No. 1 (1980), 161-163.

[37] S.POPA: "On a problem of R.V.Kadison on maximal abelian *-subalgebras," Inventiones Math. **65** (1982), 269.

[38] J.G.RATCLIFFE: "Crossed extensions," preprint.

[39] M.RIEFFEL: "Actions of finite groups on C*-algebras," Math. Scand. **47** (1980), 157-176.

[40] S.STRATILA: Modular Theory in Operator Algebras, (Editura Academiei and Abacus Press, 1981).

[41] S.STRATILA and D.VOICULESCU: "Representations of AF-algebras and of the group $U(\infty)$," Lecture Notes in Math., No. 486 (1975).

[42] S.STRATILA and L.ZSIDO: Lectures on von Neumann Algebras, (Editura Academiei and Abacus Press, 1979).

[43] C.SUTHERLAND: "Cohomology and extensions of operator algebras, II," Preprint.

[44] M.TAKESAKI: "Duality in cross products and the structure of von Neumann algebras of type III," Acta Math. **131** (1973), 249-310.

NOTATION INDEX

(Not denotes the notation section in the Introduction.)

SUBJECT INDEX